SCIENCE
A CLOSER LOOK

BUILDING SKILLS

Assessment

Mc Graw Hill **Macmillan McGraw-Hill**

Contents

Contents

Macmillan/McGraw-Hill *Assessment* in science, Grade 4, is a comprehensive program designed to familiarize students with standardized testing in science and to review the concepts covered in Macmillan/McGraw-Hill *Science: A Closer Look*. The practice tests and performance assessment activities in this book can also serve as tools in a complete program of assessment to help gauge mastery of the science content students have learned.

About This Book

The questions in this book will accustom students with standardized testing in science, including multiple-choice and open-response style questions about Life Science, Earth Science, and Physical Science, in a grade-appropriate manner. General scientific methods are stressed along with critical thinking.

The main components of this book coincide with the respective chapters and lessons in Macmillan/McGraw-Hill *Science: A Closer Look* and include:

Chapter Tests A and B: Each summative practice test covers science content from the corresponding chapters and tests students' knowledge of important vocabulary and concepts they have learned. Key concepts are tested in several ways to ensure that students comprehend core content. Skills such as making inferences, drawing conclusions, and scientific thinking are emphasized in the practice tests. Both practice tests cover the same content, but test the material in different ways, providing the teacher with several options of using the tests as pretests and posttests, chapter tests, homework assignments, or as extra practice.

Lesson Tests: These pages provide test practice and focus on specific concepts covered in each lesson of the corresponding chapter.

Performance Assessment Activity: Each activity covers a main concept from the corresponding chapter and provides students with a hands-on exercise that further reinforces the content they have learned. A rubric precedes each activity and provides guidelines for grading students' performance. Performance assessment activities require adult supervision.

How to Administer the Practice Tests

- Remove the practice test pages from the book and photocopy them for students. Answers for all questions are marked in non-reproducible blue ink.

- Separate students' desks so that students can work independently.

- Tell students that they are taking a practice test and ask them to remove everything from their desks except for several pencils. They may not speak to classmates until the test is over.

- Keep the classroom atmosphere as much like the administration of a standardized test as possible. Minimize distractions and discourage talking.

The scientific knowledge assessed in this book and in Macmillan/McGraw-Hill *Science: A Closer Look* will help students build a strong foundation in science and lay the groundwork for future learning.

Kingdoms of Life

Write the word that best completes each sentence in the spaces below. Words may be used only once.

cell	kingdom	reproduction	spores
epidermis	organisms	respiration	
germination	oxygen	seed	

1. A(n) _____kingdom_____ is the largest group into which an organism can be classified.

2. A part of the air that most plants and animals need in order to live is _____oxygen_____ .

3. The smallest unit of living matter is a(n) _____cell_____ .

4. The process by which organisms make offspring is _____reproduction_____ .

5. The protective layer of a leaf that helps keep water in is called the _____epidermis_____ .

6. Plants that do not have seeds reproduce with _____spores_____ .

7. When a seed sprouts into a new plant, this process is called _____germination_____ .

8. Cells use sugars to make energy in a process called _____respiration_____ .

9. Living things that carry out the five basic life functions are called _____organisms_____ .

10. A _____seed_____ is an undeveloped plant that carries food in a protective covering.

Circle the letter of the best answer for each question.

11. Which of the following is not an example of a living thing?

    A plants

    B animals

    (C) viruses

    D cells

12. When the female and male sex cells join together to form a seed, the process is called

    A germination.

    (B) fertilization.

    C pollination.

    D transpiration.

13. What do plants use to make food in the process of photosynthesis?

    A water, oxygen, and sunlight

    (B) water, sunlight, and carbon dioxide

    C chlorophyll, water, and gases

    D sunlight, gas, and decomposers

14. Which of the following is the broadest classification into which scientists place organisms from the same kingdom?

    (A) phylum

    B class

    C family

    D genus

15. An organism's life cycle can <u>best</u> be described as

    A the moment at which a plant germinates.

    B how long the organism can be expected to live.

    C the time during which a plant needs an animal pollinator.

    (D) the stages of growth and change in its life.

© Macmillan/McGraw-Hill

Answer the following questions.

16. **Classify** Explain why fungi are grouped in their own kingdom. How are they like plants? How are they unlike plants?

Answers will vary but may include the following: Scientists once

thought fungi were a kind of plant. Then they discovered that

fungi do not make their own food. Therefore, fungi are classified

in a separate kingdom from plants.

_____

_____

17. **Communicate** Describe the traits of the organisms found in the animal kingdom.

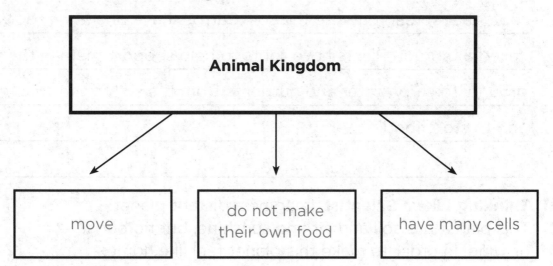

Animal Kingdom

move

do not make their own food

have many cells

18. **Observe** What are the five basic jobs all living things carry out?

All living things eat food, grow, reproduce, get rid of wastes, and

respond to their environment.

Answer the following questions.

19. What is the purpose of the tubes that run through the roots and stems of most plants?

Answers will vary but may include the following: The tubes carry

water and nutrients from the ground to the plant's leaves to help

the plant carry out photosynthesis.

_____

20. Critical Thinking Describe the root system of grass. How is it different from the root system of a carrot? What are some of the reasons these plants have roots?

Answers will vary but may include the following: Grass has fibrous

roots that spread out into the soil. Carrots have one large root,

called a taproot. Plants have roots to help support them in the

ground, to carry water and nutrients from the soil to their leaves,

and to store food.

21. Thinking Like a Scientist Imagine a distant planet. This planet has soil and air and sunshine, but no bugs or wind. In order to make this planet feel like home, seeds from fruit trees are planted. Years later, there are trees in the yard with beautiful flowers, but fruit never grows. What is this?

Answers will vary but may include the following: because this

planet has no animal pollinators or wind, fertilization cannot

happen. A seed cannot form unless pollen is transported from the

flowers' anthers to the pistils.

Circle the letter of the best answer for each question.

1. When a living thing reproduces, it creates

  A wastes.

  B pellets.

  (**C**) offspring.

  D roots.

2. Study the chart below.

Parts of Plant Cells

Name	Function
Cell wall	provides protection and support
Chloroplast	makes food
Nucleus	controls cell activities
Vacuole	?

Which function belongs in the empty box?

  A controls traits

  B creates the cell's energy

  (**C**) holds water, food, and wastes

  D gives the cell a boxlike shape

3. Animal cells cannot make their own food because they lack

  A energy.

  (**B**) chlorophyll.

  C oxygen.

  D tissues.

4. Which of the following describes a way in which viruses differ from living things?

  (**A**) Viruses cannot reproduce on their own.

  B Viruses are made of many cells.

  C Viruses are very large.

  D Viruses carry out five basic life functions.

Critical Thinking Plant cells have a thick, protective cell wall. Animal cells have only a thin cell membrane and no cell wall. Suggest a reason why plant cells need a thick protective coating.

Plants are unable to move on their own, so their cells need more

protection. Also, thick cell walls may make it easier for plants to store

food for long periods of time.

Circle the letter of the best answer for each question.

1. Which of these statements is true about protists?

   A All protists make their own food.

   B Some protists are used to bake bread.

   C Protists contain many cells.

   D All protists have a nucleus.

2. The scientific name for a common housecat is *Felis catus.* The word *Felis* refers to the animal's

   A kingdom.

   B phylum.

   C genus.

   D species.

3. Which of these is the smallest group into which an organism can be classified?

   A phylum

   B species

   C class

   D order

4. Study the diagram below.

   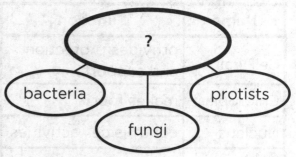

   Which title belongs in the empty oval?

   A Animal Kingdom

   B Microorganisms

   C Ancient Bacteria

   D Diseases

Critical Thinking How might some types of bacteria be helpful to the environment?

Some types of bacteria break down dead plant matter for food.

If it were not for these bacteria, dead plants would pile up in the

environment and crowd out live plants.

© Macmillan/McGraw-Hill

Circle the letter of the best answer for each question.

1. Carbon dioxide enters a plant's leaves through the

   A pigments.

   B stomata.

   C veins.

   D root hairs.

2. A tree trunk is a type of

   A root.

   B leaf.

   C stem.

   D seed.

3. What do seeds and spores have in common?

   A They both create new plants.

   B They both come from ferns.

   C They are both a type of fruit.

   D They are both found inside fruit.

4. Study the diagram below.

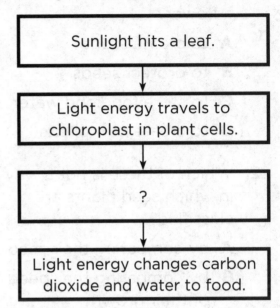

Which process belongs in the empty box?

   A Seeds create new plants.

   B Plants give off oxygen.

   C Plant cells create their own nutrients.

   D Chlorophyll traps the Sun's energy.

Critical Thinking Imagine that a plant lost all of its leaves. What do you think might happen to the plant?

The plant would die because it would have no way to make food.

Plants make food through photosynthesis, which takes place in the

leaves. If a plant did not have leaves, photosynthesis could not occur.

_____

Circle the letter of the best answer for each question.

1. What is the main purpose of a flower?

   A to create scent

   B to protect seeds

   C to store food and water

   (D) to create new plants

2. Which of these is <u>not</u> a way in which seed plants are classified?

   A by comparing their seeds

   (B) by comparing the speed of their growth

   C by comparing their roots, stems, and leaves

   D by comparing ways in which they store seeds

3. The resin of a female pinecone

   A attracts animals.

   (B) catches pollen.

   C makes flowers.

   D releases pollen.

4. Study the diagram below.

   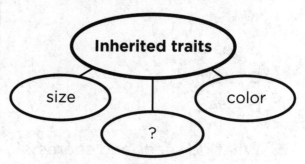

   Which detail belongs in the empty oval?

   A life cycle

   B fruit

   (C) shape

   D conifer

Critical Thinking Imagine that you are standing under a pine tree when several pinecones fall to the ground. What do you think happened when the pinecones hit the ground? What do you think will happen next?

Possible answer: When the pinecones fell on the ground, they dropped

the seeds that were inside them. The seeds will form seedlings that will

become new pine trees.

8   **Chapter 1 •** Kingdoms of Life
Assessment   Use with **Lesson 4**
How Seed Plants Reproduce

© Macmillan/McGraw-Hill

Kingdoms of Life

Write the word that best completes each sentence in the spaces below. Words may be used only once.

cell	kingdom	reproduction	spores
epidermis	organisms	respiration	
germination	oxygen	seed	

1. Red blood cells carry _____oxygen_____ through the body.

2. A(n) _____seed_____ needs water, warm temperatures, and stored food to begin to grow.

3. Organisms make offspring through the process of _____reproduction_____ .

4. A(n) _____cell_____ is the basic building block of all living things.

5. All members of the plant _____kingdom_____ have many cells, make their own food, and cannot move.

6. The _____epidermis_____ helps to keep water inside a leaf.

7. In the process of _____respiration_____ organisms use stored food to make energy.

8. In the process of _____germination_____ a seed grows into a new plant.

9. Ferns and mosses reproduce by making _____spores_____ .

10. Some _____organisms_____ are single celled, while others are organized into tissues, organs, and organ systems.

Circle the letter of the best answer for each question.

11. Water, sunlight, and carbon dioxide are all needed for which of the following processes?

   A fertilization

   B germination

   C reproduction

   (D) photosynthesis

12. A wolf and a dog belong to the same genus, but to different

   A phyla.

   (B) species.

   C families.

   D classes.

13. All organisms experience different stages of growth and change during

   (A) their life cycles.

   B reproduction.

   C their classification systems.

   D fertilization.

14. The production of a seed through the joining of female and male sex cells is a process called

   (A) fertilization.

   B germination.

   C transpiration.

   D pollination.

15. Which of the following is an example of a living thing?

   A rock

   (B) fungus

   C crystal

   D soil

Name _____ Date _____

Answer the following questions.

16. **Classify** Describe fungi and how they are classified.

Fungi seem to be plantlike, but they do not make their own food.

Therefore, fungi are classified into their own kingdom, which is

separate from the plant kingdom.

_____

_____

_____

17. **Observe** What might happen to an organism if it
stopped carrying out just one of its five basic life
functions?

The organism would stop living and eventually die.

_____

_____

18. **Communicate** Fill in the chart below. Include traits
found in the members of the animal kingdom. Use
lions as the example.

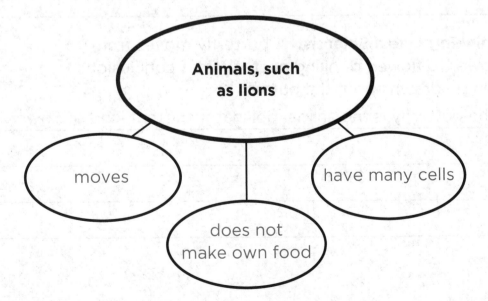

Answer the following questions.

19. What would happen if there were no tubes in the roots and stems of plants?

The plant would get no water or nutrients from the soil and would

be unable to carry out photosynthesis. The plant would die.

_____

_____

_____

_____

20. Critical Thinking Which kind of plant would better survive a drought: a potato or a clover? Why?

The potato plant would survive for a longer period during a

drought, because it has a taproot that stores food. Clover, a type

of grass, has a fibrous root system that stores very little food.

_____

_____

_____

21. Thinking Like a Scientist A butterfly moves from flower to flower, drinking nectar. What conclusions can be drawn about the butterfly?

The butterfly is the animal pollinator for that kind of flower.

_____

_____

_____

Invent a Flowering Plant

Objective: Students will build a flower of their invention. Students will recognize that every flower's job is to reproduce. Therefore, students will create appropriate anther and pistil for the flowers and be able to explain how pollination and fertilization take place.

Scoring Rubric

4 points Student creates a flower with anther and pistil. Student describes the flower and its various parts clearly and correctly. Student clearly answers the questions in Analyze the Results. Answers to all of the questions are correct.

3 points Student creates a flower that has some of the necessary parts for reproduction. Student describes the flower and its parts clearly and correctly. Student's answers to questions are partially correct.

2 points Student creates a flower without the necessary parts for reproduction. Description of the flower is partially correct. Answers to the questions are incorrect.

1 point Student attempts the assignment but the flower and the answers to the questions are mostly incorrect.

Materials

- pipe cleaners
- modeling clay
- construction paper
- egg cartons
- scissors
- glue
- tissue paper
- aluminum foil
- string
- thread
- drinking straws

Invent a Flowering Plant

Make a Model

Invent a flower using the materials provided by your teacher. Remember that your flower has to survive and reproduce, just as all living things do. Think about the characteristics of a flower before you start to build your own. What does it need to carry out the basic life function of reproduction?

Analyze the Results

1. Describe your flower. Make sure to explain the uses for each of its parts.

   Answers will vary but may include the following: All flowers
   _____

   should have an anther and a pistil.
   _____

   _____

   _____

   _____

   _____

2. How will your flower reproduce?

   Answers will vary but must include a means for pollen to be
   _____

   transferred from anther to pistil, such as by an animal pollinator
   _____

   or the wind.
   _____

The Animal Kingdom

Write the word or words that best complete each sentence in the spaces below. Words may be used only once.

birds	invertebrates	muscular system	vertebrates
cold-blooded	life cycle	nervous system	
endoskeleton	life span	sponge	

1. Animals without a backbone are ___invertebrates___ .

2. A ___sponge___ is the simplest kind of invertebrate, and its body does not have symmetry.

3. Animals with a backbone are ___vertebrates___ .

4. Animals, such as snakes, that warm their bodies in the Sun are ___cold-blooded___ .

5. The job of the ___muscular system___ is to move a body's bones.

6. The ___nervous system___ is made up of the brain, spinal cord, nerves, and sense organs.

7. A(n) ___life cycle___ is an organism's birth, growth, reproduction, and death.

8. A(n) ___life span___ is how long an organism is expected to live.

9. Animals that have hollow bones are ___birds___ .

10. A(n) ___endoskeleton___ is an internal supporting structure.

Chapter Test A

Name ______________________ Date ____________

Circle the letter of the best answer for each question.

11. Which of the following is an example of a vertebrate?

A cnidarian

B mollusk

C sponge

D amphibian

12. A shark is an example of which type of fish?

A cartilaginous

B jawless

C bony

D warm-blooded

13. Which of the following systems helps to put oxygen in blood cells and remove waste gas?

A the circulatory system

B the respiratory system

C the muscular system

D the digestive system

14. Which of the following reproduces through regeneration?

A sea star

B dragonfly

C arachnid

D squirrel

15. Which is an example of a cnidarian?

A nematode

B earthworm

C praying mantis

D jellyfish

Answer the following questions.

16. **Classify** What are the two main groups into which animals are divided?

   Animals are divided into vertebrates and invertebrates.
   _____

   _____

17. **Infer** What might happen to a vertebrate if its muscular system stopped working?

   If a vertebrate animal's muscular system stopped working, it
   _____

   would no longer be able to use its bones or move its body parts.
   _____

   The animal would die.
   _____

   _____

18. **Interpret Data** Fill in the chart below with the five stages of a butterfly's complete metamorphosis.

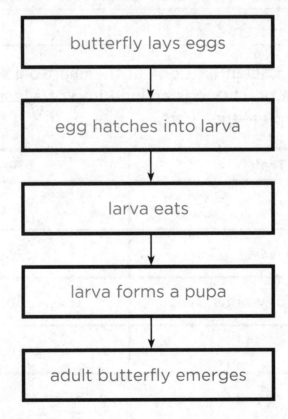

butterfly lays eggs

↓

egg hatches into larva

↓

larva eats

↓

larva forms a pupa

↓

adult butterfly emerges

Answer the following questions.

19. How are fish, amphibians, reptiles, and birds alike? How are they different?

    Answers will vary but may include: Similarities: Fish, amphibians,

    reptiles, and birds are all vertebrate animals. They all produce

    eggs. Differences: Fish are the only animals that must live in the

    water. Amphibians live in the water and on land. Reptiles live on

    land. Birds are the only animals that can fly.

20. **Critical Thinking** Why is an animal's nervous system important?

    Answers will vary but may include: The nervous system controls

    all of the other systems of the body. Therefore, it is essential to

    the proper functioning of the respiratory, circulatory, excretory,

    and digestive systems.

21. **Thinking Like a Scientist** Look at the chart below. Decide if the action listed is an inherited or a learned behavior. Complete the chart.

Trait	Behavior
a dog barking	inherited
a boy throwing a ball	learned
a bird looking for worms	learned
a baby crying	inherited

© Macmillan/McGraw-Hill

Circle the letter of the best answer for each question.

1. Clams and oysters are types of

   A arthropods.

   B cnidarians.

   (**C**) mollusks.

   D shells.

2. Which of these arthropods has the most body sections?

   A an insect

   (**B**) a centipede

   C a crustacean

   D an arachnid

3. The word *symmetry* refers to a body's

   A size.

   B color.

   (**C**) shape.

   D energy.

4. Study the diagram below.

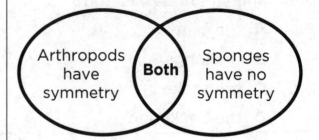

Which of these belongs in the "Both" portion of the diagram?

   A Hard outer shell

   (**B**) No backbone

   C Cannot move on their own

   D Have tentacles

Critical Thinking Why is it important for an animal such as a bird to have symmetry, but not important for an animal such as a sponge to have symmetry?

Symmetry helps to keep a bird balanced so it can walk and fly

successfully. A sponge stays in one place, so it does not need

symmetry to keep it balanced.

_____

Circle the letter of the best answer for each question.

1. Which of these is part of an animal's endoskeleton?

   A the skin

   B the lungs

   C the scales

   (D) the backbone

2. Which of these statements about amphibians and reptiles is correct?

   A Amphibians have a strong waterproof covering and reptiles do not.

   B Amphibians and reptiles must both spend time in water.

   C Amphibians are cold-blooded and reptiles are warm-blooded.

   (D) Amphibians have moist skin and reptiles have dry skin.

3. **The Life of a Frog**

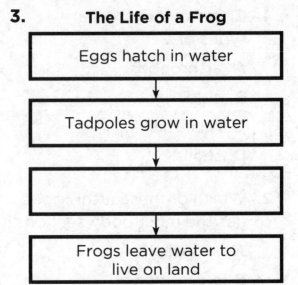

   Which belongs in the empty box?

   (A) Tadpoles turn into frogs

   B Tadpoles reproduce

   C Frogs shed skin

   D Tadpoles grow scales

4. Hollow bones help birds to

   (A) fly.

   B lay eggs.

   C stay warm.

   D breathe.

Critical Thinking Many birds can fly, but some cannot. How are the traits of a flightless bird different from those of a bird that can fly?

A flightless bird might not have hollow bones because it does not need

to lift itself up into the air. The flightless bird might also have weaker

wing muscles and shorter wings than those of a bird that can fly.

Circle the letter of the best answer for each question.

1. A skeletal system is made up of

   (**A**) bones.

   B muscles.

   C nerves.

   D blood vessels.

2. Which of these creatures has the <u>least</u> complicated nervous system?

   A a bird

   B a horse

   C a hamster

   (**D**) a worm

3. Which of these is something that is <u>not</u> carried in the blood?

   A food

   B carbon dioxide

   (**C**) digestive juices

   D oxygen

4. Study the diagram below.

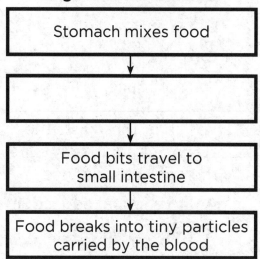

Digestion in Mammals

> Stomach mixes food
>
> ↓
>
> [ ]
>
> ↓
>
> Food bits travel to small intestine
>
> ↓
>
> Food breaks into tiny particles carried by the blood

Which detail belongs in the empty box?

A Cells turn food into particles

B Wastes exit the body

(**C**) Digestive juices break down food

D Gasses are removed from cells

Critical Thinking Imagine that a mammal was born with the nervous system of a sponge. How might this affect the way the mammal lived?

Possible answer: Mammals have complicated nervous systems. The nervous system of a sponge is very simple and has very few nerves. If a mammal had the nervous system of a sponge, it would have a hard time finding food and avoiding danger in the wild.

Name _____ Date _____

Circle the letter of the best answer for each question.

1. Molting allows an insect to

   A eat.

   B fly.

   C grow.

   D reproduce.

2. Which of these lists the stages of a butterfly's life in the correct sequence?

   A larva, pupa, egg, adult

   B pupa, larva, egg, adult

   C larva, egg, pupa, adult

   D egg, larva, pupa, adult

3. Which of these creatures reproduces by budding?

   A a hydra

   B a canary

   C a sea star

   D a butterfly

4. Study the diagram below.

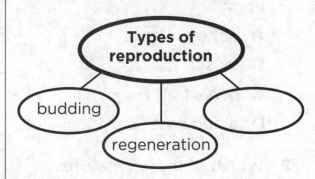

Which detail belongs in the blank oval?

   A fertilization

   B respiration

   C metamorphosis

   D molting

Critical Thinking Is sneezing an inherited behavior or a learned behavior? Explain your reasoning.

Sneezing is an inherited behavior because it is a simple reflex similar to

blinking. People do not need to be taught how to sneeze; it is a reflex

they are born with.

_____

The Animal Kingdom

**Write the word or words that best complete each sentence
in the spaces below. Words may be used only once.**

birds	invertebrates	muscular system	vertebrates
cold-blooded	life cycle	nervous system	
endoskeleton	life span	sponge	

1. Amphibians, reptiles, birds, and mammals are all
   different examples of _____vertebrates_____ .

2. The five senses (sight, hearing, taste, touch, and smell)
   are all part of the _____nervous system_____ .

3. The largest animal group is made up of _____invertebrates_____ .

4. The skeletal system and the _____muscular system_____ work
   together to move the body's bones.

5. In a typical _____life cycle_____ , an organism is born,
   grows, reproduces, and dies.

6. One of the simplest invertebrates is the _____sponge_____ .

7. Warm-blooded vertebrates with hollow bones are
   called _____birds_____ .

8. Animals, such as lizards, that warm their bodies in the
   Sun during the day are _____cold-blooded_____ .

9. The internal supporting structure of some animals is
   called its _____endoskeleton_____ .

10. How long an organism can be expected to live is
    called its _____life span_____ .

Circle the letter of the best answer for each question.

11. What does the circulatory system do?

   A It puts oxygen in blood cells and removes waste gas.

   (B) It moves blood through the body.

   C It breaks down food and removes it from the body.

   D It removes wastes from the body.

12. How does a penguin reproduce?

   A through external fertilization

   (B) through the development and hatching of eggs

   C through the regeneration of one of its body parts

   D through a bud that develops and then breaks off the parent

13. Which of the following is an example of an invertebrate?

   A bullfrog

   (B) echinoderm

   C jawless fish

   D snake

14. What do a shark and a goldfish have in common?

   (A) a backbone

   B no jaw or teeth

   C a skeleton composed of cartilage

   D their size

15. Budding is a process in which an animal reproduces

   A when its body grows into a new organism.

   B by the fertilization of an egg cell with a sperm cell.

   (C) by growing offspring on its body that later break off.

   D by cloning itself when it divides in half.

Answer the following questions.

16. **Classify** Describe the feature that is used to group the animal kingdom into two main groups. What are these groups called?

    Answers will vary but may include: Animals are divided into the

    two main groups—vertebrate and invertebrate. The feature that is

    used to group them is whether or not they have a backbone.

17. **Infer** What might happen to a vertebrate animal if its digestive system stopped working?

    Answers will vary but may include: If a vertebrate animal's

    digestive system stopped working, it would no longer be able to

    digest food for energy. The animal would die.

18. **Interpret Data** What process is shown in the pictures below? Label each of the five stages.

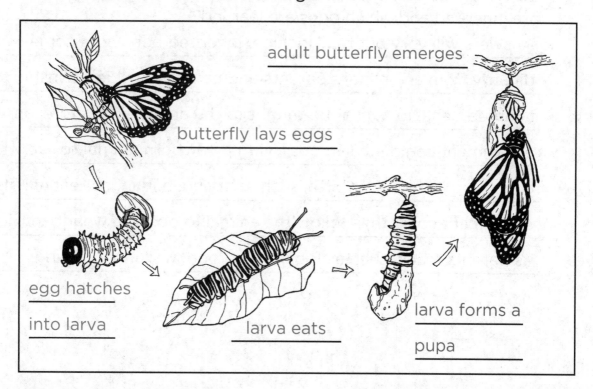

adult butterfly emerges

butterfly lays eggs

egg hatches
into larva

larva eats

larva forms a
pupa

Answer the following questions.

19. What are two similarities among mammals, reptiles, and birds? What are some differences?

Answers will vary but may include: Mammals, reptiles, and birds are

all vertebrate animals. They all must breathe air and most of them

live on land. Birds are the only ones that can fly. Some mammals live

only in the water, while most reptiles and birds live on land.

20. Critical Thinking What function does an animal's excretory system perform?

The excretory system removes waste products from the body.

21. Thinking Like a Scientist How do inherited behaviors compare with learned behaviors? How would you design an experiment to determine what behaviors are inherited and which ones are learned?

Answers will vary but may include: Inherited behaviors come

directly from an animal's parents and can be considered instincts.

Learned behaviors must be taught to the animal. Therefore, an

experiment could involve observing a newborn animal to see what

behaviors it exhibits at birth, such as crying, barking, or eating. The

experiment could then follow the animal to observe what behaviors

its parents teach, such as looking for food, walking, or flying.

Life Cycle

Objective: Students will create a mobile to represent the stages of an animal's life cycle. Students may choose any animal. Students will recognize that a life cycle involves at least the following steps: birth, growth, reproduction, death. Students who choose an animal that undergoes metamorphosis must include those steps as well.

Materials

- construction paper
- markers
- scissors
- pencils
- glue

Scoring Rubric

4 points Student creates a mobile of an animal's life cycle that includes birth, growth, reproduction, and death. If the animal undergoes metamorphosis, the mobile includes those steps as well. Student clearly answers the questions in Analyze the Results. Answers to all of the questions are correct.

3 points Student creates a mobile of an animal's life cycle that includes most of these stages: birth, growth, reproduction, and death. If the animal undergoes metamorphosis, the mobile includes most of those steps as well. Student's answers to questions are partially correct.

2 points Student creates a life cycle that includes fewer than half of the stages: birth, growth, reproduction, death. If the animal undergoes metamorphosis, the mobile includes fewer than half of those steps. Answers to the questions are incorrect.

1 point Student attempts the assignment but the life cycle and the answers to the questions are incorrect.

Life Cycle

Make a Mobile

Make a mobile to show the stages of an animal's life cycle. You may choose any animal. Remember that the animal you choose must undergo birth, growth, reproduction, and death. If you choose an animal that undergoes metamorphosis, make sure to include those steps in the life cycle as well. Think about the steps in the animal's life cycle before you start to make your mobile. Make sure you show pictures and write descriptions to represent each stage of your animal's life cycle.

Analyze the Results

1. Describe the animal you chose. Make sure to list all of the steps in the animal's life cycle.

   Answers will vary but must include birth, growth, reproduction, _____

   and death. _____

   _____

   _____

   _____

2. How does your animal reproduce? Describe the process.

   Answers will vary but may describe one of the following processes:

   budding, regeneration, or fertilization of an egg with a sperm cell.

   _____

   _____

© Macmillan/McGraw-Hill

Exploring Ecosystems

Write the word that best completes each sentence in the spaces below. Words may be used only once.

abiotic factors	deciduous forests	food chain	tundra
biomes	decomposers	grassland	
consumers	ecosystem	producers	

1. Energy passes from one organism to another in a
   ___food chain___ .

2. A(n) ___ecosystem___ can be small, such as a log, or large, such as a forest.

3. Trees in ___deciduous forests___ have broad leaves that block sunlight.

4. Worms, bacteria, and fungi are ___decomposers___ that break down dead organisms.

5. Water and rocks are ___abiotic factors___ in an environment.

6. A prairie, which has very few trees, is one type of
   ___grassland___ .

7. Animals that eat other organisms to get the energy they need to survive are ___consumers___ .

8. Plants found in the ___tundra___ include arctic moss, arctic willow, and bearberry.

9. There are six major ___biomes___ on Earth.

10. The source of all of the food in an ecosystem comes from ___producers___ .

Circle the letter of the best answer for each question.

11. Animals that eat only plants are

    A carnivores.

    (B) herbivores.

    C omnivores.

    D producers.

12. Which of the following is an example of an abiotic factor in an ecosystem?

    A worms

    B mosses

    (C) rocks

    D trees

13. What is the largest biome in the world?

    A deciduous forest

    B grassland

    C tropical rain forest

    (D) taiga

14. Oaks, maples, and hickories are examples of the trees that grow in a

    A tropical rain forest biome.

    B tundra biome.

    (C) deciduous forest biome.

    D taiga biome.

15. Which of the following shows the direction in which energy flows in a food chain?

    (A) producer → consumer → decomposer

    B consumer → producer → carnivore

    C carnivore → herbivore → decomposer

    D producer → carnivore → herbivore

Answer the following questions.

16. Interpret Data Fill in the summary box below with the kind of ecosystem described in the chart.

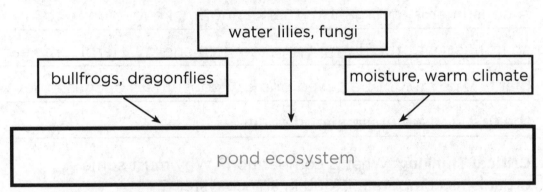

17. Classify Describe a grassland.

Grasslands have mild climates with good soil for farming. They get

some rainfall and may have fires in hot, dry summer seasons.

_____

_____

18. Predict Make an ocean food web. Show where shrimp, fish, squid, and whales fit into the web.

Ocean Food Web

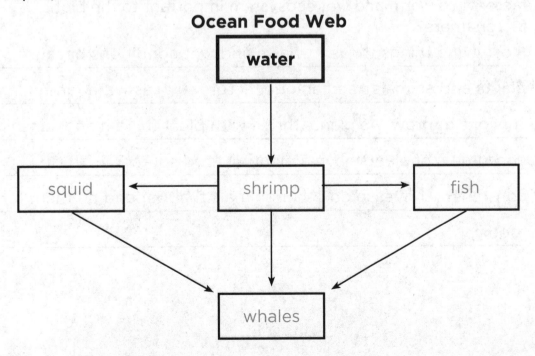

Answer the following questions.

19. What is a consumer? What do herbivores consume? Why might herbivores be called <u>primary</u> consumers?

A consumer is an organism that cannot make its own food, and

so it must eat other living things to gain energy. Herbivores eat

plants. They might be called primary consumers because they are

the first consumers in a food chain.

20. Critical Thinking What is competition? Why must some organisms compete for food in an ecosystem?

Competition is when more than one organism in an ecosystem

depend on the same food source. Some organisms must compete

for food if there is not enough food in the ecosystem.

21. Thinking Like a Scientist Ecosystems that are warm and wet usually have more plant and animal life than cold and dry ecosystems. Why might this be the case? How are the abiotic factors in a warm and wet ecosystem important to the biotic factors there?

Most living things thrive in warm and wet conditions because

plants and animals need abiotic factors such as water and

sunlight to grow. As plants thrive with plentiful rain and sunshine,

consumers of plants have a plentiful food source. As these

consumers thrive, so do other animals that depend on them for

food.

Circle the letter of the best answer for each question.

1. Which of these is true about an ecosystem?

  A Only one population at a time lives in an ecosystem.

  B An ecosystem contains populations, but no communities.

  C An ecosystem can be a home to many different populations.

  D An ecosystem is made up of only abiotic factors.

2. The study of ecosystems is known as

  A economics.

  B zoology.

  C biology.

  D ecology.

3. Study the chart below.

Biotic Factors	Abiotic Factors
pine trees	light
squirrels	water
bacteria	

Which of these belongs in the blank box?

  A soil

  B ferns

  C worms

  D cacti

4. Butterflies and daisies can be members of the same

  A population.

  B species.

  C community.

  D family.

Critical Thinking Imagine that a golf course is built in a spot that was once a forested area. Many trees are cut down to make way for golf greens. Human-made ponds are installed where no ponds had existed before. How do you think this would affect the area's ecosystem?

Many animals would leave because the trees they depend on for food

and shelter are gone. Some new animals might come to the area that

had not lived there before. The ponds might bring frogs and other

animals, such as ducks, to the area.

Circle the letter of the best answer for each question.

1. A biome is a type of

   A grassland.

   B wetland.

   C population.

   (D) ecosystem.

2. Grass, shrubs, and trees are the three main types of plants that grow in a

   A desert.

   (B) savannah.

   C tundra.

   D taiga.

3. Most animals in the tundra are likely to have

   A webbed feet.

   B feathers.

   (C) thick coats.

   D fins.

4. Study the diagram below.

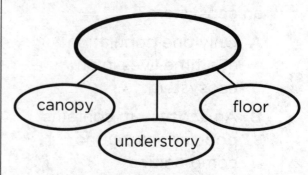

   Which of these belongs in the blank oval?

   A Parts of a Grassland

   B Parts of a Prairie

   (C) Parts of a Forest

   D Parts of a Tundra

Critical Thinking Animals live in ecosystems that suit their needs and abilities. How might fish that live in a very deep ocean ecosystem be different from fish that live close to the ocean's surface?

Possible answer: Fish that live very deep in the ocean would be able to

survive in very cold water and swim with little or no light. Fish that live

close to the surface would depend on sunlight to keep them warm and

to help them see.

© Macmillan/McGraw-Hill

Circle the letter of the best answer for each question.

1. Producers could not make food without

   A omnivores.

   B worms.

   (C) sunlight.

   D competition.

2. Which of these statements is true?

   A Insects receive very little of the Sun's energy through their food.

   (B) A lion receives less of the Sun's energy in its food than a caterpillar does.

   C Animals and insects capture energy directly from the Sun.

   D Plants capture very little of the Sun's energy.

3. Study the diagram below.

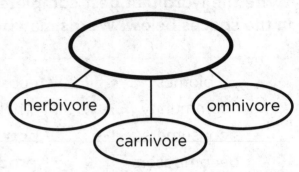

   Which of these belongs in the blank oval?

   A Producers

   B Decomposers

   C Food Chain

   (D) Consumers

4. Nutrients are added to the soil with the help of

   (A) decomposers.

   B herbivores.

   C predators.

   D prey.

Critical Thinking Tall trees and small plants compete for sunlight and nutrients. Describe some advantages that tall trees might have over small plants.

Tall trees can soak up more sunlight because they are closer to the

Sun. They also have large root systems to absorb water and nutrients

from the soil.

© Macmillan/McGraw-Hill

Name _____ Date _____

Exploring Ecosystems

Write the word that best completes each sentence in the spaces below. Words may be used only once.

biomes	ecosystem	tropical rain forests
community	habitat	tundra
consumers	population	
decomposers	producers	

1. Plants, algae, and other organisms that use energy from the Sun to make food are ____producers____ .

2. Each organism in an ecosystem has its own special ____habitat____ , or place where it lives.

3. A(n) ____ecosystem____ is made up of both the living and nonliving things in an environment.

4. Carnivores and herbivores are ____consumers____ .

5. Large ecosystems on Earth are called ____biomes____ .

6. An ecosystem's ____community____ includes all of the populations in that ecosystem.

7. A(n) ____tundra____ is a dry, treeless biome where the ground is frozen year-round.

8. A(n) ____population____ is all the members of a species in an ecosystem.

9. Organisms that break down dead organisms are ____decomposers____ .

10. Large amounts of rainfall are found in ____tropical rain forests____ .

Circle the letter of the best answer for each question.

11. The tundra is the only biome in which

 A there is less than 10 inches of rain per year.

 B the leaves on trees change color in the fall.

 C there are 24 hours of sunlight in the summers.

 (D) the ground is frozen all year long.

12. Carnivores are animals that

 A eat only plants.

 B produce their own food.

 (C) eat other animals.

 D consume plants and animals.

13. What is the role of water, soil, sunshine, and rocks in an ecosystem?

 A They are all biotic factors.

 B They are all eaten by animals in the ecosystem.

 (C) They are all abiotic factors.

 D They are all producers of their own food.

14. Which are always found at the base of an energy pyramid?

 A consumers

 B carnivores

 (C) producers

 D decomposers

15. Which of the following is a sandy or rocky biome with little rainfall?

 A taiga

 (B) desert

 C tundra

 D grassland

Answer the following questions.

16. **Classify** Fill in the chart below with the details of the ecosystems found in the tropical rain forest.

Tropical Rain Forest	Details
canopy	top of trees; home to many animals
understory	hot and damp; animals travel here for food
forest floor	very dark; lots of bugs and snakes

17. **Predict** What might happen to a food web if the organism in the center of the web completely disappeared?

If the organism in the center of a food web disappeared, the

organisms that depend on that one would need to find alternate

food sources.

_____

_____

_____

18. **Interpret Data** Fill in the details of the chart below with the types of biotic and abiotic factors found in a desert ecosystem.

Type of Ecosystem	Biotic and Abiotic Factors
desert	cactus
	sand
	hot and dry climate
	lizards and scorpions

© Macmillan/McGraw-Hill

Name _____ Date _____

Answer the following questions.

19. What do carnivores consume? Why might carnivores be called <u>secondary</u> consumers?

Carnivores eat other animals. They might be called secondary

consumers because they are the second consumers in a food

chain.

20. Critical Thinking Look at the food web below. What is the relationship between the rabbit and the snail?

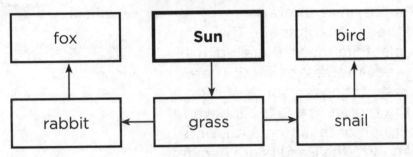

Meadow Food Web

The rabbit and the snail both eat grass, and so they compete for

the same food source.

21. Thinking Like a Scientist Do the following things make up an ecosystem? Why or why not? If not, what could be added or taken away to form an ecosystem?

> caterpillars, ferns, butterflies, spiders, foxes,
> grass, snails, rabbits, birds

Answers will vary but may include: No, these things do not make

up an ecosystem. An ecosystem requires abiotic factors including:

water, sunlight, a warm climate, rocks, and soil.

© Macmillan/McGraw-Hill

Exploring Ecosystems

Biome Diorama

Objective: Students will build a shoebox diorama of one of the following biomes: grassland, deciduous forest, tropical rain forest, tundra, taiga, or desert. The diorama will accurately represent the landscape of the region, and include a label and a short description of the biome.

Materials

- shoebox
- colored clay
- assorted rocks
- construction paper
- popsicle sticks
- markers
- scissors
- tape

Scoring Rubric

4 points Student builds an accurate representation of the selected biome. The diorama includes plant life found there, such as grasses, trees, flowers, mosses, or cacti. There may be small representations of some of the animals found in the region. There is a label and an accurate description of the biome. Answers to the questions in Analyze the Results are accurate and complete.

3 points Student builds a representation of the selected biome. The diorama includes plant life found there and some animals. There is a label and a description of the biome. Answers to the questions in Analyze the Results are mainly accurate and complete.

2 points Student builds a representation of the selected biome, but it is not complete. The diorama includes some of the plant life found there but no animals. There is no label or description of the biome. Answers to the questions in Analyze the Results are mainly inaccurate and incomplete.

1 point Student attempts the assignment but the diorama and the answers to the questions are mostly incorrect.

Biome Diorama

Make a Model

A diorama is a small model of a scene or landscape. Using the materials provided by your teacher, design and make a shoebox diorama of a biome of your choosing. Whether you make a grassland, tropical rain forest, deciduous forest, taiga, tundra, or desert, make sure to include representations of the biome's major features. Label the diorama and include a description of what plants and animals make this area their home.

Analyze the Results

1. In what parts of the world is your biome found? What abiotic factors play a major role there?

   Answers will vary depending on the biome chosen. Students

   should accurately identify where in the world the biome is located,

   and include a description of at least two abiotic factors, such as

   sunlight and water.

   _____

2. What kinds of ecosystems can be found in this biome? What are the major animals found there?

   Answers will vary depending on the biome chosen. Students

   should accurately identify one or two ecosystems common in this

   biome, and include a description of at least two animals that make

   this place their home.

   _____

Surviving in Ecosystems

Write the word that best completes each sentence in the spaces below. Words may be used only once.

accommodation	deforestation	hibernate	tropisms
adaptations	endangered	mimicry	
camouflage	extinct	stimulus	

1. Some animals ___*hibernate*___ to avoid the cold of winter.

2. When one living thing looks like another living thing, it is called ___*mimicry*___ .

3. A(n) ___*stimulus*___ is something in an environment that causes a living thing to react.

4. Traits that help an organism survive in its environment are called ___*adaptations*___ .

5. A(n) ___*accommodation*___ is an organism's response to change.

6. A species becomes ___*extinct*___ when it dies out.

7. The responses of plants to light, water, and gravity are ___*tropisms*___ .

8. Animals that blend in with their environment have an adaptation called ___*camouflage*___ .

9. Organisms are ___*endangered*___ when there are very few of their species alive.

10. When trees and shrubs are cut down by ___*deforestation*___ , ecosystems are destroyed.

Circle the letter of the best answer for each question.

11. Which of the following is an example of mimicry?

    A a giraffe has a long neck to reach high leaves

    B an insect's body is shaped like a thorn

    C a lizard changes color to blend in with its environment

    D a skunk uses a strong odor to scare away predators

12. Plants respond to stimuli by

    A adapting to their environment.

    B moving to a different location.

    C changing their pattern of growth.

    D growing different types of roots.

13. After a flood, a mouse starts to eat food it normally would not eat. This is an example of

    A mimicry.

    B camouflage.

    C tropism.

    D accommodation.

14. Why are pandas endangered?

    A people destroyed most of their food supply

    B they cannot protect themselves from predators

    C they are unable to adapt to their environment

    D they cannot learn how to use camouflage for protection

15. Which of the following is not an example of a camel's adaptation to desert life?

    A nostrils that can close to keep out sand

    B humps that store fat for energy when food is scarce

    C wide hooves to walk on sand

    D camels migrate to cooler climates in the summer

Answer the following questions.

16. **Predict** What changes might occur in a forest ecosystem after a large forest fire?

Trees and plants will be destroyed, but decaying matter will help

to enrich the soil and help new plants grow from seeds. Also, the

population of some animals will decrease because of the loss of

food and shelter. Other animals, like deer, might change what they

eat in order to survive.

17. **Interpret Data** Fill in the chart below with the responses most plants have to the following stimuli.

Stimulus	⟶	Response
light	⟶	turns and grows towards it
gravity	⟶	roots grow down and stems grow up
water	⟶	roots grow towards it

18. **Communicate** Explain how camouflage can help an animal survive in its environment. Give one example of how an animal uses camouflage.

An animal uses camouflage to blend into its environment. This

helps the animal hide from its predators and sneak up on its prey.

An arctic fox's coat turns white in the winter to blend in with the

snow.

© Macmillan/McGraw-Hill

Answer the following questions.

19. How do plants adapt to cold winters?

Some plants in colder climates lose their leaves each winter, so

the leaves do not die from the cold. To save energy, some plants

do not photosynthesize during winter.

20. Critical Thinking How is a brightly colored flower on a plant an example of an adaptation?

The bright flower will attract animal pollinators. The plant adapted

to attract animal pollinators in order to reproduce and thus

survive.

21. Thinking Like a Scientist An ecosystem is severely flooded. What can people do to help save this ecosystem?

Once the flood waters recede, people could plant trees and other

plants in the region. People could volunteer with animal rescue

organizations to help save the animals and bring them back to

their homes when the land is dry.

Circle the letter of the best answer for each question.

1. A giraffe's long neck is an example of

   A mimicry.

   B an adaptation.

   C hibernation.

   D camouflage.

2. Which of these is an example of camouflage?

   A a moth that looks and acts like a hummingbird

   B a bear that sleeps in a cave all winter long

   C a colorful fish that lives among brightly colored coral

   D an elephant that uses its large ears to fan itself

3. Study the chart below.

Animal	Adaptation
polar bear	has thick fur
great horned owl	has night vision
duck	

   Which of these belongs in the blank box?

   A imitates other animals

   B has webbed feet

   C lives in ponds

   D eats bread

4. When a bird flies south for the winter, it

   A hibernates.

   B camouflages.

   C migrates.

   D mimics.

Critical Thinking Adaptations are traits that help living things survive in their environment. Describe an adaptation that has helped humans to survive.

Possible answer: Humans can think and solve problems. This has

helped humans develop cures for diseases, for instance, and avoid

danger.

Name _____ Date _____

Circle the letter of the best answer for each question.

1. Which would <u>not</u> change a potted plant's growth pattern?

   A receiving less water

   B receiving less sunlight

   C rotating the pot

   (D) exposing it to a strong smell

2. Which of these is <u>not</u> an example of an adaptation?

   A a cactus has a thick, waxy cover to hold water in

   B a tree loses its leaves in the winter

   C a lily grows bright red flowers to attract pollinators

   (D) a plant has green leaves

3. What is a tropism?

   (A) a plant's response to its environment

   B a change in climate

   C something found in soil

   D an animal's adaptation

4. Which of these belongs in the blank box?

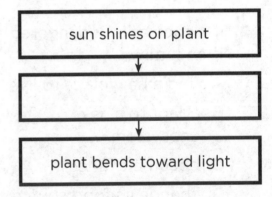

   A gravity pulls plant upward

   (B) a chemical reacts in plant

   C plant begins to flower

   D bees pollinate plant

Critical Thinking The Venus's-flytrap, has the ability to trap and eat insects. What might this adaptation tell you about its environment?

Possible answer: It suggests that there are many insects in its

environment and that it is hard to get nutrients from other sources.

_____

_____

© Macmillan/McGraw-Hill

Circle the letter of the best answer for each question.

1. What is an accommodation?

A an environment's response to its climate

B the buildup of pollution in an ecosystem

C a community's response to overpopulation

(D) an individual organism's response to change

2. If a species is endangered, then it may

A find a new place to live.

B adapt to a new ecosystem.

(C) become extinct.

D change its food source.

3. Study the diagram below.

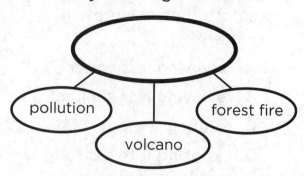

Which of these belongs in the blank oval?

A Types of Pollution

B Effects of Overpopulation

C Causes of Deforestation

(D) Events That Change an Ecosystem

4. Which is <u>least likely</u> to occur when an ecosystem changes?

(A) All of the organisms will adapt to the change.

B Some animals will leave.

C Some organisms will die.

D The land may benefit.

Critical Thinking How might small plants benefit from deforestation?

Possible answer: Tall trees compete with small plants for light and nutrients; therefore, small plants would probably thrive because they would have less competition.

Surviving in Ecosystems

**Write the word that best completes each sentence
in the spaces below. Words may be used only once.**

accommodation	deforestation	hibernate	tropism
adaptations	endangered	mimicry	
camouflage	extinct	stimulus	

1. The brown pelican is an example of an
   _____endangered_____ animal.

2. A plant reacts to a(n) _____stimulus_____ by
   changing its direction and pattern of growth.

3. By copying the traits of the bumblebee, the hoverfly
   uses _____mimicry_____ to protect itself.

4. The arctic fox changes color with the season, using
   _____camouflage_____ to survive.

5. When people cut down entire forests for their own
   needs, it is called _____deforestation_____ .

6. When animals _____hibernate_____ , they lie very
   still for a long period of time and use very little energy.

7. When living things change their habits and behaviors,
   this is called _____accommodation_____ .

8. Plants become _____extinct_____ when they
   cease to exist.

9. A plant bending to reach sunlight is an example of
   a(n) _____tropism_____ .

10. Plants in the desert have _____adaptations_____ for
    saving water.

Circle the letter of the best answer for each question.

11. Which of the following is an example of a desert plant adaptation?

   A does not photosynthesize year-round

   B loses its leaves in winter

   C has long, woody trunk

   (D) has spongy tissue to retain water

12. An arctic hare's seasonal change of fur color is an example of

   A mimicry.

   B hibernation.

   C tropism.

   (D) camouflage.

13. Plants have tropisms to all of the following except

   A light.

   B water.

   C chemicals.

   (D) odors.

14. Dinosaurs became extinct due to environmental factors of nature, while endangered animals today are mostly the result of

   A being killed by their natural predators.

   (B) the actions and behavior of people.

   C sudden changes in the weather.

   D their natural ability to adapt.

15. Which of the following is not an example of an animal's adaptation to desert life?

   A a fennec fox has large ears to give off heat

   (B) a snake sheds its skin so it can grow larger

   C a kangaroo rat gets water from food so it does not need to drink

   D a camel stores fat in its hump for energy when food is scarce

© Macmillan/McGraw-Hill

Answer the following questions.

16. **Communicate** Describe an adaptation that helps a snow leopard survive.

A snow leopard has spots so that it blends into its environment.

Camouflage helps the snow leopard because it can hide from

other animals and it can sneak up on its prey more easily.

17. **Predict** What changes might occur in a waterfront ecosystem after a large tidal wave?

Possible answer: Trees and plants will be destroyed by flooding;

however, new plants will begin to grow from seeds and from

more sunlight once the land dries. Also, the population of some

animals will decrease because of the loss of food and shelter.

Other animals might make accommodations to survive in their new

environment.

18. **Interpret Data** Fill in the chart below describing animals and their adaptations.

Animal	Adaptation
arctic hare	fur color changes to match season
bear	hibernates during winter
hedgehog	protects itself with spines
skunk	protects itself with a strong odor
hoverfly	looks like a bee
kangaroo rat	gets water from its food

Answer the following questions.

19. What are three similarities between desert plants and plants from other climates?

Desert plants and plants from milder climates all have root

systems, leaves, and many of them have flowers to attract animal

pollinators. Plants in all environments make accommodations or

adaptations in order to survive in their environments.

20. Critical Thinking How is the spongy tissue inside a desert plant an example of an adaptation?

The spongy tissue acts like a sponge to absorb and hold water, so

that the plant can survive long periods of time with limited rainfall.

21. Thinking Like a Scientist A new power plant is set to open soon. What dangers could a local pond face?

Possible answer: The pond might be in danger of being

polluted. Land might be destroyed to clear area for the power

plant.

Surviving in Ecosystems

Adapt to It!

Objective: Students will choose a plant or animal to place in an unfamiliar ecosystem. They will then invent an adaptation for the plant or animal that will allow it to survive in its new home.

Scoring Rubric

4 points Student creates an adaptation appropriate for the new ecosystem. Student describes the chosen plant or animal, the adaptation, and the new ecosystem clearly and correctly. Answers to the questions in Analyze the Results are clear and correct.

3 points Student creates an adaptation which is somewhat appropriate for the new ecosystem. Student describes the chosen plant or animal, the adaptation, and the new ecosystem clearly and correctly. Answers to the questions in Analyze the Results are partially correct.

2 points Student creates an adaptation which is somewhat appropriate for the new ecosystem. Descriptions of the chosen plant or animal, the adaptation, and the new ecosystem are partially correct. Answers to the questions in Analyze the Results are mainly incorrect.

1 point Student attempts the assignment but the adaptation and the answers to the questions are mostly incorrect.

Materials

- pipe cleaners
- modeling clay
- scissors
- glue
- construction paper
- tissue paper
- aluminum foil
- string
- thread
- drinking straws

© Macmillan/McGraw-Hill

Adapt to It!

Make a Model

Select any plant or animal you like, and then decide what it would need to survive in unfamiliar surroundings. Then, choose from the materials provided by your teacher to create a model of the plant or animal that you chose. Create and add a new adaptation that would allow it to live in a different ecosystem than its own. Think about the characteristics of the new ecosystem before you begin.

Analyze the Results

1. What plant or animal did you decide to adapt? What new ecosystem will this plant or animal now have as its home? Describe the new ecosystem.

   Answers will vary. The plant or animal mentioned should not

   normally live in the ecosystem described.

   _____

2. Describe the adaptation you gave to the plant or animal. How will this adaptation allow the plant or animal to survive in its new home?

   Answers will vary. Any reasonable adaptation should be accepted

   if the means of survival are explained correctly.

   _____

3. Would this adaptation help or harm the plant or animal you chose in its natural ecosystem? Describe its natural ecosystem and explain why this adaptation would help or harm the organism if it were to return there.

   Answers will vary. Any logical answer should be accepted.

   _____

   _____

Shaping Earth

Write the word that best completes each sentence in the spaces below. Words may be used only once.

crust	landslide	mountain	volcanoes
flood	mantle	terminus	
fold	moraines	tornado	

1. The layer of rock that lies below Earth's crust is called the _____mantle_____ .

2. A _____mountain_____ is a tall landform that rises to a peak.

3. A _____tornado_____ is a column of air that spins rapidly and moves across the ground in a narrow path.

4. The outermost layer of Earth is called the _____crust_____ .

5. A _____fold_____ is a bend in the rock layers that happens when plates slowly push into each other.

6. After heavy rainfall, the land can become filled with water and a _____flood_____ can occur.

7. Rock debris accumulates at the _____terminus_____, or downhill end, of a glacier.

8. The mounds of rock debris deposited by a glacier are called _____moraines_____ .

9. The Hawaiian Islands are famous for their _____volcanoes_____, which can sometimes erupt.

10. When heavy rains fall on a mountainside, a _____landslide_____ can occur.

Name _____ Date _____

Circle the letter of the best answer for each question.

11. Rock is broken down and moved to another place during the process of

   A erosion and deposition.

   (B) weathering and erosion.

   C physical and chemical weathering.

   D weathering and deposition.

12. Freezing and thawing are a type of

   A chemical weathering.

   B erosion.

   (C) physical weathering.

   D deposition.

13. The surface of Earth is rapidly changed by all of the following except

   A landslides.

   (B) glaciers.

   C earthquakes.

   D volcanoes.

14. Each of these are examples of Earth's underwater features except

   (A) the outer core.

   B the continental shelf.

   C the continental slope.

   D the ocean ridge.

15. Which of the following natural disasters is not caused by storms?

   A landslide

   (B) earthquake

   C tornado

   D flood

Answer the following questions.

16. **Infer** Why do earthquakes form along faults?

Answers will vary but may include the following: Earth's crust is made of moving plates. As these plates get pushed and pulled the crust moves along cracks, called faults. When the energy built up along a fault is released, it causes the crust to shift, which causes an earthquake.

17. **Experiment** Design an experiment to determine the effect plants have on soil erosion. Explain your answer.

Answers will vary but may include the following: To see the effects of plants on soil erosion, I could plant grass in one tray and fill another tray with only soil. Then I could tilt the trays, run water down the slopes, and observe the effects. I think that the tray with the grass would have less erosion than the tray with only dirt because the plants would help keep the dirt from running off. The tray with only soil would most likely lose much of its soil in the water runoff.

Answer the following questions.

18. Describe how physical and chemical weathering affect rocks. How are both types of weathering alike and how are they different? Record your answers in the Venn Diagram.

Physical Weathering **Both** **Chemical Weathering**

does not change chemical makeup of rocks

breaks rocks down

changes chemical makeup of rocks

19. **Critical Thinking** There have been very heavy rains all along the Mississippi River. What might happen to the land surrounding the Mississippi?

Answers will vary but may include the following: There would be

severe floods throughout the region because the land around the

Mississippi is a drainage basin for the Mississippi River.

20. **Thinking Like a Scientist** As rivers near the ocean, they slow down. What kind of landform can one expect to see along coastlines where rivers empty into the ocean? How could one test this hypothesis?

Answers will vary but may include the following: The landform

where rivers empty into the ocean are deltas. To test this

hypothesis, I could use books and maps to locate coastlines

with rivers and study their landforms along the rivers' mouths to

determine if these landforms are deltas.

© Macmillan/McGraw-Hill

Circle the letter of the best answer for each question.

1. A delta is made up of

    A melted iron.

    B sharp rocks.

    C gullies.

    (D) sand.

2. Where can underwater canyons form?

    (A) continental slope

    B continental rise

    C ocean ridges

    D drainage basins

3. Earth's crust is made up of

    A water.

    (B) rock.

    C iron.

    D sand.

4. Study the illustration below.

    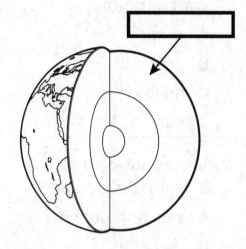

    Which of these belongs in the blank box?

    A outer core

    B inner core

    (C) mantle

    D crust

Critical Thinking How might ocean waves affect a beach over thousands of years? Explain your answer.

Answers will vary but may include: Moving water changes the shape of

the land around it, so ocean waves would change the shape of a beach.

The waves might even wear away the beach, making it smaller.

_____

Circle the letter of the best answer for each question.

1. A plateau is a tall landform with a top that is

   A round.

   B pointed.

   C hollow.

   (**D**) flat.

2. Vibrations caused by earthquakes are known as

   A seismographs.

   B tsunamis.

   (**C**) seismic waves.

   D eruptions.

3. Which of these **best** describes what Earth's plates are made up of?

   A broken pieces of Earth's crust

   (**B**) Earth's mantle and crust

   C hardened lava

   D melted rock

4. Study the diagram below.

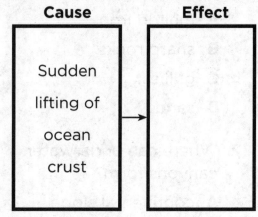

Cause	Effect
Sudden lifting of ocean crust	

   Which of these belongs in the "Effect" box?

   A Volcano

   B Fault

   C ocean ridge

   (**D**) Tsunami

Critical Thinking Some earthquakes are very strong while others are barely noticeable. Why?

When plates push against each other, energy builds up in the rock.

When the rock breaks, it releases this energy, which shifts the crust.

When a very large amount of stored energy is released, it creates a

strong earthquake, while a small amount of released energy produces a

weak earthquake.

Circle the letter of the best answer for each question.

1. Which of these does <u>not</u> contribute to physical weathering?

    (**A**) chemical reactions

    B growing plants

    C flowing water

    D freezing water

2. Large rocks left behind by glaciers are known as

    A erosion.

    B glacial till.

    C moraines.

    (**D**) glacial debris.

3. Which of these contributes to the formation of sand dunes?

    (**A**) deposition

    B erosion

    C weathering

    D terminus

4. Study the diagram below.

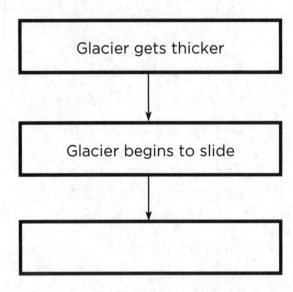

    Which belongs in the empty box?

    A Snow collects quickly

    B Pressure builds on the mound

    (**C**) Glacier tears at the ground

    D Glacier bottom turns to ice

Critical Thinking How might crashing waves on a beach affect the ocean floor over thousands of years?

Answers will vary but may include: When waves crash on a beach, they

carry sand and small rocks away from it. The sand and small rocks may

drop to the ocean floor and eventually change the shape of the ocean

floor.

Circle the letter of the best answer for each question.

1. Where is a hurricane <u>most likely</u> to form?

   (A) over a warm ocean

   B over flat land

   C in a forested area

   D over a mountain range

2. The sudden movement of snow and ice down a mountain is called a(n)

   A landslide.

   B hurricane.

   C flood.

   (D) avalanche.

3. A landslide is something that occurs

   A slowly, over many years.

   B only in coastal regions.

   (C) rapidly, without much warning.

   D on land that is mostly flat.

4. A tornado is a(n)

   A swirling water storm.

   (B) narrow column of wind.

   C overflow of water onto land.

   D common cause of falling rocks.

Critical Thinking Scientists believe the temperature of Earth is increasing. How might this affect the weather over time? Explain your answer.

Answers will vary but may include: If Earth's temperature increases,

then there might be more hurricanes. Scientists think that hurricanes

are becoming more common because of these higher temperatures.

Rising temperatures could also cause glaciers to melt, which could

cause flooding.

Shaping Earth

Write the word that best completes each sentence in the spaces below. Words may be used only once.

crust	landslide	mountains	volcanoes
flood	mantle	terminus	
fold	moraines	tornado	

1. The ____crust____ is made of solid rock and forms the outermost layer of Earth.

2. Below Earth's crust is a layer of rock called the ____mantle____ .

3. Landforms that have tall peaks are ____mountains____ .

4. When plates slowly push into each other at the edges of two continents, a ____fold____ results.

5. Structures that look similar to mountains, but can sometimes erupt with hot lava are called ____volcanoes____ .

6. A glacier drops much of its rock debris at its ____terminus____, or downhill end.

7. Glaciers push small rocks, boulders, and dirt into mounds called ____moraines____ .

8. A ____flood____ is caused by an overflow of water.

9. Heavy rains in a mountain region can cause a ____landslide____, sending rocks and soil swiftly downhill.

10. A column of whirling wind that moves across the ground and causes destruction is called a ____tornado____ .

Circle the letter of the best answer for each question.

11. Which of the following is a process that causes a rock to break apart without changing the rock type?

   A deposition

   (B) physical weathering

   C thawing

   D chemical weathering

12. Earth's surface is slowly changed by which of the following natural events?

   (A) glaciers

   B earthquakes

   C landslides

   D avalanches

13. Weathering and erosion are the processes through which rock is

   A physically and chemically changed.

   B chemically altered and moved to a new location.

   (C) broken down and moved to another place.

   D chemically altered and broken into smooth pebbles.

14. How do earthquakes differ from other natural disasters, such as tornados and floods?

   A They are caused by fast winds.

   B They are a result of too much rain.

   (C) They are not caused by storms.

   D They are unrelated to plate movement.

15. The continental shelf and the continental slope are examples of

   A landforms.

   B Earth's layers.

   C mountains.

   (D) underwater features.

Answer the following questions.

16. Infer Draw a picture of what happens just under Earth's surface during an earthquake. Label the different parts of the model.

> Drawing should include an understanding of the following
>
> concept: There are breaks in Earth's crust where the plates
>
> come together. Faults are where rocks move along the cracks.
>
> These are the areas where the plates push together
>
> or slide apart, which causes earthquakes.

17. Experiment Look at the picture below and hypothesize about what is happening. Design an experiment that would test this hypothesis.

Hypothesis: This hillside is experiencing erosion. To test this

hypothesis, I would plant more trees and plants in the region and

observe the effects on the hillside's soil.

Answer the following questions.

18. In what ways are physical and chemical weathering alike? In what ways are they different? Fill in the chart below.

Physical Weathering	Similarities	Chemical Weathering
Rocks break down into smaller parts without changing the rocks' minerals.	Both cause rocks to break down into smaller parts.	Rocks break down by changing the chemical makeup of the rocks' minerals.

19. Critical Thinking Describe the term drainage basin. What is the result of heavy rains on a river's drainage basin?

A drainage basin is the area of land around a river, over which

water from rain or melting snow runs downhill and into the river.

Drainage basins can experience severe flooding during heavy

rains.

20. Thinking Like a Scientist Deltas form along some coastlines. What causes the formation of these deltas?

Rivers slow down as they come closer to the ocean, and begin to

deposit silt and soil into the water. Therefore, deltas form along

coastlines.

_____

_____

© Macmillan/McGraw-Hill

Landform Poster

Objective: Students will create a poster drawing of two landforms of their choice. The illustrations will accurately represent these landforms and will include labels for the various parts of the landforms.

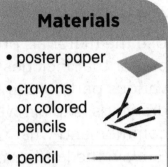

Materials

- poster paper
- crayons or colored pencils
- pencil

Scoring Rubric

4 points Student illustrates two landforms of his or her choosing. Student labels each landform correctly. Student writes a clear and accurate description of what these two landforms have in common. Student explains each landform to another student accurately and clearly. Student clearly explains his or her answers to the questions in Analyze the Results.

3 points Student illustrates two landforms of his or her choosing. Student labels most parts of the landforms correctly. Student writes an accurate description of what these two landforms have in common, but the description is unclear. Student is able to explain most of the parts of the landforms to another student accurately and clearly. Answers to the questions in Analyze the Results are mostly correct, but contain a few errors.

2 points Student illustrates one landform, or illustrates two landforms but labels most parts incorrectly. Student writes a description of what the two landforms have in common that is not accurate for the most part. Student is able to explain a few details of the landforms to another student accurately and clearly. Answers to the questions in Analyze the Results are mostly incorrect.

1 point Student does not accurately illustrate a landform or correctly label any of its parts. Student writes an inaccurate description of what the two landforms have in common. Student is not able to explain any of the parts of the landforms to another student accurately or clearly. Student does not answer the questions in Analyze the Results.

Name _____ Date _____

Landform Poster

Communicate

Use the materials provided by your teacher to create a poster that includes two kinds of landforms. Label the various parts of the landforms you choose. Beneath your drawings, explain what these two landforms have in common and what differences there are between them. Explain the landforms to another student.

Analyze the Results

1. What are the differences between the two landforms in your poster?

   Answers will vary but may include the following depending on

   which landforms were chosen: Mountains rise high above the

   surrounding land, while valleys are low areas of land between hills

   or mountains.

2. What processes are involved in the formation of the landforms in your poster?

   Weathering, erosion, and deposition are involved in the formation

   of all landforms.

3. Explain how weathering will affect the landforms in your poster.

   Answers will vary but may include the following: Landform areas

   may be weathered by blowing sand, cracked by freezing water,

   tumbled in moving water, and/or changed chemically by acid rain.

Saving Earth's Resources

Write the word or words that best complete each sentence in the spaces below. Words may be used only once.

environment	groundwater	minerals	soil profile
fossil	igneous rock	pollution	
fossil fuel	metamorphic rocks	reservoir	

1. Melted rock that cools and hardens is called _____igneous rock_____ .

2. A vertical section of soil from the surface to the bedrock is a(n) _____soil profile_____ .

3. A fuel made from the remains of ancient living things is a(n) _____fossil fuel_____ .

4. The building blocks of rocks are _____minerals_____ .

5. Water stored in pore spaces and cracks in underground rocks is _____groundwater_____ .

6. The preserved evidence of a living thing from long ago is a(n) _____fossil_____ .

7. The _____environment_____ includes all the living and nonliving things in an area.

8. Rocks formed from other rocks by heat and pressure are called _____metamorphic rocks_____ .

9. A human-made lake that is used to hold fresh water is a(n) _____reservoir_____ .

10. A harmful material that has been added to an environment is called _____pollution_____ .

Circle the letter of the best answer for each question.

11. Which statement is false?

A A mineral's luster describes how light bounces off its surface.

B A diamond can scratch any other mineral.

C The mineral quartz comes in only one color.

D The color of a mineral may be different from its streak.

12. Which of the following best describes topsoil?

A It is rich in humus and minerals and it is home to many living things.

B It is light and hard and it contains bits of mineral and clay pieces.

C It is made up of coarse, broken-down bedrock and no plant roots.

D It is not permeable, so water cannot pass through it.

13. Which of the following is not a renewable resource?

A air

B water

C fossil fuel

D animals

14. Which of the following could help an urban area get safe drinking water from a different part of the state?

A a waterfall

B a lake

C a reservoir

D a watershed

15. Acid rain and other forms of air pollution can be caused by

A the overuse of fossil fuels.

B dumping garbage outside of landfills.

C the overuse of fertilizers by farmers.

D treating crops with pesticides.

© Macmillan/McGraw-Hill

Answer the following questions.

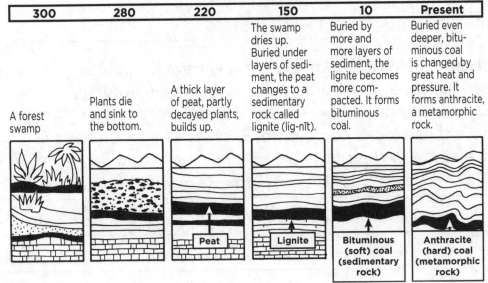

THE STORY OF COAL

Millions of Years Ago

300	280	220	150	10	Present
A forest swamp	Plants die and sink to the bottom.	A thick layer of peat, partly decayed plants, builds up.	The swamp dries up. Buried under layers of sediment, the peat changes to a sedimentary rock called lignite (lig-nīt).	Buried by more and more layers of sediment, the lignite becomes more compacted. It forms bituminous coal.	Buried even deeper, bituminous coal is changed by great heat and pressure. It forms anthracite, a metamorphic rock.
		Peat	Lignite	Bituminous (soft) coal (sedimentary rock)	Anthracite (hard) coal (metamorphic rock)

16. **Interpret Data** Look at the time line above. On the following lines, summarize how anthracite coal is formed.

Plants died and sank to the bottom of a swamp. Over time, layers

of decayed plants built up and formed peat. The peat turned into

lignite under the pressure of the sediment above it. The lignite

turned into bituminous coal. Heat and pressure turned it into

anthracite coal.

17. **Experiment** Black soil near a river is thick and wet. Sand in a desert is dry and blows away easily. Describe an experiment that would test which soil is best for farming.

Bean seeds are planted in two containers, one with river soil and

the other with desert soil. The containers are watered equally and

exposed to the same amount of sunlight. The container in which

the bean seeds sprout is the best for farming.

Answer the following questions.

18. Describe the materials that make up soil.

Soil is a mixture of weathered rock, decayed plant or animal

matter, and pockets of air or water.

19. Critical Thinking Bituminous and anthracite coal are
fossil fuels that can be burned to produce electricity
and heat. Describe how the term *fossil fuel* explains
where these rocks come from. Explain why these fuels
will eventually run out.

The term "fossil fuel" suggests that these rocks were made from

ancient plants or animals. Because it took millions of years for

these rocks to form, humankind will eventually run out of them.

They are also being burned faster than they could ever

be replaced.

20. Thinking Like a Scientist What impact could the
increased use of pesticides on crops have on the
surrounding environment?

These chemicals can pollute streams and groundwater,

which can harm plants and animals in the environment. These

chemicals can also pollute drinking water and cause illness.

Circle the letter of the best answer for each question.

1. A mineral that has a low number on the Mohs' Scale

   A is not easy to scratch.

   B is only one color.

   C is harder than a diamond.

   D is easy to scratch.

2. A diamond's shine is referred to as its

   A streak.

   B luster.

   C color.

   D reflection.

3. Which of these is true about sedimentary rocks?

   A They are made of many layers.

   B They form very quickly.

   C They are unchanged by weathering.

   D They contain many small air pockets.

4. In terms of relative age, where are the oldest layers of a sandstone rock located?

   A at the top of the rock

   B at the bottom of the rock

   C in the middle of the rock

   D at the top and bottom of the rock

Critical Thinking Ancient hunters used obsidian rock to make spears. Given this information, what can you conclude about the minerals in obsidian rock? Explain your reasoning.

Possible answer: If hunters used obsidian rock for spears, could

probably be shaped into a sharp point. This suggests that obsidian

breaks into sharp pieces, like glass.

_____

Circle the letter of the best answer for each question.

1. Which of these <u>best</u> describes humus?

   A coarse, weathered rock in the topsoil layer

   B a portion of the B horizon, or subsoil layer

   C a part of the soil that contains weathered rocks and minerals

   (D) a part of the soil that contains decayed plant or animal matter

2. The rate at which water passes through the soil is called

   A texture.

   (B) permeability.

   C weathering.

   D erosion.

3. In which soil layer would you find the <u>most</u> plant roots?

   (A) topsoil

   B subsoil

   C clay

   D bedrock

4. Study the diagram below.

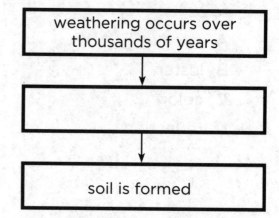

Which of these belongs in the blank box?

   A mineral bits trickle down from the topsoil

   B bedrock forms

   (C) rocks break down into smaller pieces

   D water is collected

Critical Thinking Do you think fallen leaves in the autumn affect the soil in any way? Explain your reasoning.

Possible answer: Leaves that fall in the autumn probably improve the

soil. The leaves decay and become humus, which provides the nutrients

to enrich the soil.

© Macmillan/McGraw-Hill

Name _____ Date _____

Circle the letter of the best answer for each question.

1. Which of these is <u>not</u> true of fossil fuels?

   A They are expensive to use.

   B They can be easily replaced.

   C They pollute the air.

   D They form very slowly.

2. Which of these animal parts is <u>most likely</u> to be preserved as a fossil?

   A fur

   B skin

   C teeth

   D whiskers

3. Wood that turns to stone is

   A petrified.

   B decayed.

   C renewable.

   D imprinted.

4. Study the diagram below.

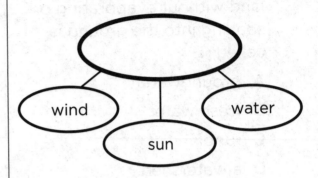

Which of these belongs in the blank oval?

   A Fossil Fuels

   B Renewable Resources

   C Nuclear Power

   D Nonrenewable Resources

Critical Thinking What kinds of things might a scientist be able to learn about an ancient creature from its foot imprint? Explain your ideas.

Possible answer: A scientist might be able to tell how much a creature

weighed depending on how deep the imprint is. An imprint showing

very sharp claws might tell scientists that the creature was a ferocious

hunter; a webbed foot would show that the creature could swim.

Circle the letter of the best answer for each question.

1. Water that flows over the land without evaporating or soaking into the ground is called

   A groundwater.

   B fresh water.

   C runoff.

   D a watershed.

2. Groundwater can be obtained by

   A digging a well.

   B collecting runoff.

   C building a reservoir.

   D collecting rainwater.

3. A reservoir

   A collects water from underground wells.

   B uses pipes to move water into cities.

   C contains water that is too salty to drink.

   D pumps water up from deep beneath the ground.

4. Which of these is not true about salt water?

   A Salt water covers almost three fourths of the Earth's surface.

   B Glaciers and ice caps are made of frozen salt water.

   C Salt water cannot be used to water crops.

   D Salt water cannot be used as drinking water.

Critical Thinking Why might a community that lives near a watershed become concerned when a chemical factory is built nearby?

Possible answer: The factory might produce harmful chemicals and

other pollutants that would be carried over the land by runoff. The

polluted runoff might contaminate the watershed.

© Macmillan/McGraw-Hill

Circle the letter of the best answer for each question.

1. Chemicals and solid wastes are two examples of

   A compost.

   (B) pollution.

   C conservation.

   D waste water.

2. One can conserve water by

   A doing laundry in the morning.

   (B) repairing a leaky garden hose.

   C using a dishwasher.

   D drinking tap water.

3. Recycling is a way to

   A create fossil fuels.

   B slow soil erosion.

   C transport waste to a landfill.

   (D) prevent pollution.

4. Farmers rotate crops in order to

   (A) conserve nutrients.

   B prevent erosion.

   C protect crops from acid rain.

   D conserve water.

Critical Thinking Many parks and other public areas have posted signs that say "Please clean up after your pets." How might cleaning up after pets help to reduce water pollution?

Possible answer: Animal waste pollutes the environment. If people do

not clean up after their pets, the animal waste might be washed into a

storm drain and into a nearby water source.

Saving Earth's Resources

Write the word or words that best complete each sentence in the spaces below. Words may be used only once.

compost	irrigation	pore spaces	soil profile
fossil fuel	metamorphic rocks	recycle	
imprint	minerals	relative age	

1. Scientists use a(n) _____soil profile_____ to study the layers of soil, called horizons, which form over time.

2. Natural, nonliving substances that make up rocks are _____minerals_____ .

3. We can _____recycle_____ materials like glass and paper.

4. Tracks, body outlines, and leaf prints are called _____imprint_____ fossils.

5. The age of one thing compared to another is its _____relative age_____ .

6. Marble and slate are common types of _____metamorphic rocks_____ .

7. The _____pore spaces_____ in soil act like filters.

8. An example of a nonrenewable resource that people use to create heat and electricity is a(n) _____fossil fuel_____ .

9. People can conserve soil by spreading _____compost_____ in their gardens.

10. In some places, _____irrigation_____ is used to bring fresh water to crops.

Circle the letter of the best answer for each question.

11. Which of the following is rich in humus and minerals and home to many living things?

    (A) topsoil

    B subsoil

    C weathered bedrock

    D metamorphic rock

12. Which of the following is a renewable resource?

    (A) soil

    B coal

    C oil

    D natural gas

13. Which of the following is not a good way to identify a mineral?

    A how it splits into pieces

    (B) by its color

    C how hard it is

    D by its streak

14. The overuse of fossil fuels leads to

    A flooding.

    (B) pollution.

    C fertile soil.

    D good crops.

15. A reservoir could supply an urban setting with

    (A) safe drinking water from a rural part of the state.

    B a good swimming spot for families.

    C a good source of wind-created energy.

    D freedom from annual river flooding.

Answer the following questions.

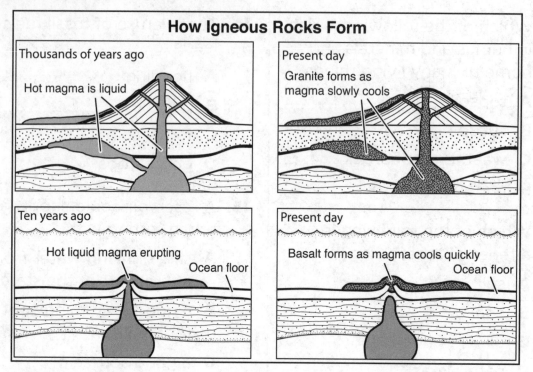

How Igneous Rocks Form

Thousands of years ago

Hot magma is liquid

Present day
Granite forms as
magma slowly cools

Ten years ago

Hot liquid magma erupting   Ocean floor

Present day

Basalt forms as magma cools quickly
Ocean floor

16. **Interpret Data** Look at the time lines above. Describe
how granite and basalt are different.

When magma turns to granite, the process is slow and the texture

of the rock is rough. When magma turns into basalt, the process is

very fast and the texture of the rock is smooth.

17. **Experiment** Gravity and rushing water remove
mineral-rich topsoil from steep slopes, leaving thin layers
of topsoil on the slopes. Rich, deep layers of topsoil build
up on flat lands. Describe an experiment that would test
which soil would be better suited for farming.

Plant corn seeds in two containers of soil—one with a thin layer

on a sloping surface and one with a thick layer on a flat surface.

Water equally and expose to the same sunlight. The container in

which the corn seeds sprout will be the best soil.

Answer the following questions.

18. What substance is made up of pieces of rock, decayed plant and animal matter, and pockets of air or water?

These substances combine to create soil.

_____

19. Critical Thinking Explain how scientists have come to understand what life looked like long ago. How does this help to explain where fossil fuels come from?

Scientists have discovered fossils of many different kinds of plants

and animals that were created as they decayed. Scientists have

learned that as these plants and animals decayed, they left behind

the materials that can be used to make fossil fuels like coal and oil.

_____

20. Thinking Like a Scientist A watershed feeds rivers and streams. There are signs posted in the watershed notifying people that they are in a watershed region. Why are these signs posted in the watershed, and what do they hope to prevent?

The effects of pollution in a watershed can be far reaching. If the

watershed is polluted, the rivers it feeds will also become polluted.

This leads to pollution in the oceans. The signs are meant to

discourage people from polluting the watershed.

Saving Earth's Resources

Make a 3 Rs Poster

Objective: Students will create a poster illustrating the 3 Rs: Reduce, Reuse, and Recycle. They will use the 3 Rs as headings and select two or more photographs, or draw pictures of common items under each heading. Next to each item they will write a sentence about how to reduce, reuse, or recycle it. Posters may be hung in the school hallways to encourage other students.

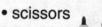

Materials

- poster board
- markers
- scissors
- glue
- photos of commonly used items that can be recycled, reduced, or reused, such as cans, plastic bottles, and sheets of paper

Scoring Rubric

4 points Student creates a poster with headings that include the 3 Rs: Reduce, Reuse, Recycle. Student includes photographs or illustrations of two or more items for each heading and writes one sentence about how to reduce, reuse, or recycle each item. Student clearly answers the questions in Analyze the Results. Answers to all of the questions are correct.

3 points Student creates a poster with headings that include the 3 Rs: Reduce, Reuse, Recycle. Student includes one item photograph or illustration for each heading. Student writes one or two words about how to reduce, reuse, or recycle each item. Student answers the questions in Analyze the Results. Answers to the questions are partially correct.

2 points Student creates a poster with headings that include the 3 Rs: Reduce, Reuse, Recycle. Student does not include photographs of items for each heading, or places items under the wrong heading. Student does not write anything about how to reduce, reuse, or recycle the items. Student attempts to answer the questions in Analyze the Results. Answers to the questions are incorrect.

1 point Student attempts the assignment but the poster is incomplete and the answers to the questions are mostly incorrect or not attempted.

Make a 3 Rs Poster

Make a Model
Make a poster about the 3 Rs; Reduce, Reuse, Recycle. Use "Reduce, Reuse, and Recycle" as a heading for your poster. Select two or more photographs or draw your own illustrations of common items that can be reduced, reused, or recycled. Next to each item, write one sentence about how to reduce its use, how to recycle the item, or how to reuse it.

Analyze the Results
1. Summarize the main idea of your poster in two or three sentences.

   Answers will vary but may include the following: People should use less of something if it cannot be recycled or reused. People should recycle or reuse everything else. This will help keep the environment clean and safe.

2. Your poster described three things people can do to help keep the environment free from pollution. What are some other things people can do to help the environment?

   Answers will vary but may include the following: riding bikes or walking instead of driving cars; composting at home and at school; picking up litter; building eco-friendly homes and buildings.

Weather and Climate

Write the word or words that best complete each sentence in the spaces below. Words may be used only once.

air masses	clouds	rain gauge	wind vane
barometer	cold front	thermometer	
climate	current	warm front	

1. A tool called a(n) ___barometer___ measures air pressure.

2. The seasonal weather pattern that happens year after year is called ___climate___ .

3. A(n) ___cold front___ forms when cold air pushes under warm air.

4. Tiny water droplets or ice crystals form ___clouds___ .

5. A(n) ___wind vane___ points in the direction from which the wind is blowing.

6. A directed flow of a gas or a liquid is a(n) ___current___ .

7. A tube that collects water and measures how much rain has fallen is a(n) ___rain gauge___ .

8. When warm air pushes over cold air a(n) ___warm front___ forms.

9. Large areas of air that share the same properties are called ___air masses___ .

10. A(n) ___thermometer___ measures temperatures in degrees Celsius or degrees Fahrenheit.

Circle the letter of the best answer for each question.

11. Humidity is low

(A) in deserts.

B over oceans.

C over lakes.

D in rain forests.

12. Evaporation is the process in which

A water vapor cools and changes into a liquid.

B water falls from clouds down to the Earth.

(C) water becomes a gas.

D water cools quickly and turns into solid ice.

13. What usually happens at the boundary between two air masses?

(A) A front is formed and the weather changes.

B The cold air is pushed down and the temperature drops.

C A stationary front forms and it rains for days.

D The warm air is pushed up and the temperature rises.

14. Which two factors most strongly affect a region's climate?

A longitude and ocean currents

B precipitation and closeness to a mountain

C wind and temperature

(D) latitude and closeness to an ocean or lake

15. Earth's water

A is mostly in its vapor state at any given time.

(B) has been recycled for billions of years.

C freezes as it rises into the atmosphere.

D only falls to Earth as rain.

Answer the following questions.

Use the information in this table to answer questions 16 and 17.

How Do Weather and Climate Affect Crops?

Temperature	Rainfall	Crop	Rate of Growth
40°	steady	potato	good
40°	light	rice	poor
60°	light	potato	very poor
60°	light	rice	very poor
90°	heavy	potato	poor
90°	heavy	rice	good

16. **Interpret Data** What is the relationship between temperature, rainfall, and rate of growth for potatoes? What is it for rice?

   Potatoes need steady rainfall and cool temperatures in order to

   grow well. Rice needs heavy rainfall and hot temperatures to grow

   well.

17. **Infer** Would it be possible to grow potatoes and rice on the same farm? Why or why not?

   No. Potatoes grow best in cold places with steady rainfall, and rice

   grows best in hot, wet climates.

18. **Predict** The climate on one side of a mountain is dry. What will be the climate on the other side? Explain why.

   Answers will vary but may include: The climate will most likely be

   wet. As a cloud moves up a mountain the water vapor condenses

   and falls as precipitation. By the time it passes over the mountain

   it will be dry.

Answer the following questions.

19. **Critical Thinking** Why might people be very uncomfortable if a stationary front was forecasted for their region?

   Answers will vary but may include: A stationary front is a

   boundary between air masses that does not move. Rainfall, high

   humidity, and a lack of a breeze could make people feel very

   uncomfortable.

   _____

20. **Thinking Like a Scientist** The atmosphere is made up of oxygen, nitrogen, carbon dioxide, and water vapor. What are some of the possible effects if the gases in the atmosphere changed? What would happen if the atmosphere could no longer hold water vapor?

   Answers will vary but may include: Changes to Earth's atmosphere

   would be deadly. If there was no oxygen, people would not be

   able to breathe. If there was no carbon dioxide, plants would die.

   If the atmosphere could no longer hold water vapor, a key part of

   the water cycle would be affected and Earth's water would run

   out.

Name _____ Date _____

Circle the letter of the best answer for each question.

1. Earth's atmosphere is mainly made up of

   A oxygen and carbon dioxide.

   B carbon dioxide and nitrogen.

   (C) nitrogen and oxygen.

   D nitrogen and helium.

2. The stratosphere is located

   (A) directly above the troposphere.

   B directly below the troposphere.

   C directly above the mesosphere.

   D directly above the thermosphere.

3. Study the chart below.

Tool	Purpose
barometer	measures air pressure
hygrometer	measures humidity
anemometer	

   Which of these belongs in the blank box?

   A measures air pressure

   (B) measures wind speed

   C measures rainfall

   D measures air temperature

4. Sleet and hail are types of

   A gases in the atmosphere.

   B layers in the atmosphere.

   C air pressure.

   (D) precipitation.

Critical Thinking In a two-story house, which floor do you think would be the warmest—the lower level or the upper level? Use what you know about hot and cold air to explain your reasoning.

The upper level would be the warmest. Warm air rises because its

particles are far apart and it weighs less. Cool air has particles that are

close together and weigh more. Warm air is more likely to rise upstairs

and cool air is more likely to sink downstairs.

© Macmillan/McGraw-Hill

Name _____ Date _____

Circle the letter of the best answer for each question.

1. Which of these is true about water vapor?

   (A) It is water in a gas state.

   B It is water in a solid state.

   C It is water in a liquid state.

   D It is not measurable.

2. What is the energy source that drives the water cycle?

   A cloud formation

   (B) the Sun

   C lakes and streams

   D condensation

3. During the water cycle, water moves

   A only between clouds.

   B between lakes and soil.

   C between plants and soil.

   (D) between Earth and the air.

4. Study the diagram below.

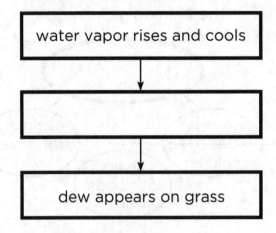

   Which of these belongs in the blank box?

   A liquid evaporates

   (B) gas changes to liquid

   C ice forms

   D runoff flows over the ground

Critical Thinking What factors do you think might cause a cumulus cloud to change into a cumulonimbus cloud? What causes the cumulonimbus cloud to appear dark?

Cumulus clouds can change depending on how high they are in Earth's

atmosphere, and the size and type of water droplets they contain. As

moisture builds up in cumulus clouds, they begin to appear darker

and thicker and become cumulonimbus clouds. These clouds cause

precipitation.

Circle the letter of the best answer for each question.

1. Which of these belongs in the blank oval?

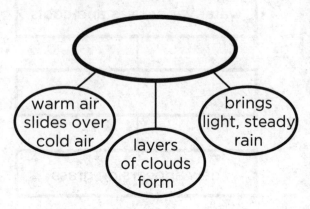

   warm air slides over cold air

   layers of clouds form

   brings light, steady rain

   A Hurricane

   B Cold Front

   C Warm Front

   D Tornado

2. Thunder occurs when

   A air masses collide.

   B lightning heats air quickly.

   C warm air hits the ground.

   D a warm front forms.

3. When does a cold front form?

   A when a cold air mass pushes under a warm air mass

   B when a warm air mass pushes under a cold air mass

   C when two cold air masses collide

   D when the boundary between two air masses stops moving

4. A hurricane forms

   A over dry land.

   B over cold ocean waters.

   C over warm lakes and streams.

   D over warm ocean waters.

Critical Thinking The local weather forecast says there will be a cold front moving through your area tomorrow that may become a stationary front. Is it necessary to water your garden this evening? Why?

Possible answer: I would not water my plants in the garden this

evening. Cold fronts often bring stormy weather. If the front becomes

stationary it could rain for days.

_____

Circle the letter of the best answer for each question.

1. Global winds move air

   A between the equator and the poles.

   B around the equator.

   C from the ground to the air.

   D from land to the ocean.

2. Which of these <u>best</u> describes a tropical climate?

   A cold and windy

   B warm and dry

   C hot and humid

   D cool and rainy

3. Which of these items would <u>not</u> have a current?

   A air

   B ocean

   C plain

   D river

4.

   | cloud moves up mountain |

   ↓

   | precipitation falls |

   ↓

   | air mass passes over mountain and air is dry |

   ↓

   | each side of mountain has a different climate |

   Which does the chart describe?

   A global winds

   B the mountain effect

   C circular air currents

   D water currents along lines of latitude

Critical Thinking In terms of latitude, where on the globe would you expect to find tropical rain forests? Explain your answer.

Rain forests would be located near the equator because areas near the

equator are warm and rainy.

Weather and Climate

Write the word or words that best complete each sentence in the spaces below. Words may be used only once.

atmosphere	cold front	thermometer	water vapor
barometer	currents	warm front	
clouds	forecast	water cycle	

1. Satellites help us _____ forecast _____ the weather.

2. Cumulus, stratus, and cirrus are all types of _____ clouds _____ .

3. The liquid in a(n) _____ thermometer _____ rises and falls with the temperature.

4. A(n) _____ cold front _____ often moves quickly and brings stormy weather.

5. The weight of the air pushing on an area can be measured with a(n) _____ barometer _____ .

6. Water moves between Earth's surface and atmosphere by evaporation, condensation, and precipitation in the _____ water cycle _____ .

7. Some _____ currents _____ in the ocean transport warm water from the equator to the poles.

8. The blanket of air that covers the Earth is the _____ atmosphere _____ .

9. A(n) _____ warm front _____ often brings light, steady rain and causes the temperature to rise after the rain passes.

10. An invisible part of the air around us is called _____ water vapor _____ .

Circle the letter of the best answer for each question.

11. Which is <u>not</u> a form of precipitation?

    A snow

    B sleet

    C humidity

    D hail

12. Circular patterns in the oceans are formed by

    A currents.

    B air masses.

    C rainfall.

    D humidity.

13. The greater the water vapor in the air, the greater the

    A air temperature.

    B humidity.

    C air pressure.

    D wind speed.

14. Condensation is the process in which

    A water falls from clouds down to the Earth.

    B water vapor cools and changes into a liquid.

    C water cools quickly and turns into solid ice.

    D water vapor rises into the atmosphere.

15. A moving cold air mass is

    A a stationary front.

    B precipitation.

    C a cold front.

    D air pressure.

Name _____   Date _____

Answer the following questions.

Use the information in this graph to answer the following questions.

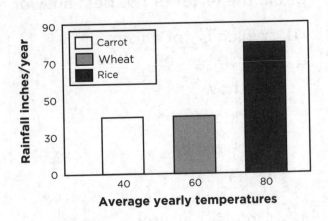

16. **Interpret Data** What does this bar graph show about rainfall and temperature requirements for the three crops?

Carrots need steady rainfall and cool temperatures in order to

grow well. Rice needs heavy rainfall and hot temperatures. Wheat

needs warm temperatures and steady rainfall.

17. **Infer** What would happen to a wheat crop planted in a hot, dry place? What is the best location to grow wheat?

The wheat crop would probably not grow very well in a hot,

dry place. It should be planted somewhere with moderate

temperatures and steady rainfall.

18. **Predict** Could all three crops be grown on the same farm? Why or why not? What possible combinations of crops could there be on a farm with moderate temperatures and steady rainfall?

There is no farm that could grow all three of these crops, but one

farm could grow wheat and carrots at different times of year.

Carrots could be grown in the fall or winter, and wheat in the

spring or summer.

Answer the following questions.

19. **Critical Thinking** How does weather forecasting affect people's lives? What could happen if we did not have weather forecasting technology?

Answers will vary but may include: Weather forecasting helps

people plan for bad weather, such as tornadoes or hurricanes. If

we did not have weather forecasting, people would not have time

to find shelter when storms were approaching.

_____

_____

20. **Thinking Like a Scientist** Suppose a scientist tracking a thunderstorm notices a swirling mass of air beginning to form. What might happen? What should the scientist do?

Answers will vary but may include: The swirling air mass may form

a tornado. The scientist should alert the government and tell the

public to take shelter.

_____

_____

_____

Name _____ Date _____

Weather and Climate

Make a Weather Map

Objective: Students will make a weather map of any region of the world. The map must show a weather forecast that is possible for that region. The map should be labeled and should have a key, so people reading it can understand the symbols used on the map. Maps should be hung around the classroom so students can view each other's maps.

Scoring Rubric

4 points Student creates an accurate weather map of a selected region of the world. The map shows what weather will likely occur in that region. The map is labeled and has a key, so others can read it easily. Answers to the questions in Analyze the Results are accurate and complete.

3 points Student creates a weather map of a selected region of the world. The map shows what weather will likely occur in that region. The map has a key, but may not be labeled correctly. Answers to the questions in Analyze the Results are mostly accurate and complete.

2 points Student creates a weather map of a selected region of the world. The map is missing a label and a key, so it is confusing to others trying to read it. Answers to the questions in Analyze the Results are mostly inaccurate or incomplete.

1 point Student attempts the assignment but the weather map and the answers to the questions are mostly incorrect.

Materials

- blank outline maps of different regions of the world

- colored pencils, markers, crayons

- index cards for cutouts

- scissors

- tape and glue

Make a Weather Map

Make a Model

Use the materials provided by your teacher to make a weather map. Your map can be of any region of the world, but the weather coming through the region must be accurate. Make sure to include labels or a key, so that people reading your map know what the symbols on your map mean. Hang the maps on the walls of your classroom and study the other maps to see what is happening around the world today.

Analyze the Results

1. Describe your map. What region of the world does it cover? What weather might this region be experiencing today?

   Answers will vary but must include an accurate description of the

   chosen region of the world and its weather.

2. How did you discover what kind of weather normally affects this region? Have you visited this region, or read about it in books? Did you have to research the region's weather? If so, what tools did you use to do your research?

   Answers will vary but must include a description of how the

   student discovered the type of weather normally experienced in

   the region chosen for the map.

The Solar System and Beyond

Write the word or words that best complete each sentence in the spaces below. Words may be used only once.

asteroids	lunar eclipse	rotation	telescope
axis	phases	solar system	
constellations	revolution	stars	

1. The apparent shapes of the Moon in the sky are called its _____ phases _____ .

2. One object travels around another in a(n) _____ revolution _____ .

3. A(n) _____ axis _____ is a real or imaginary line that an object spins around.

4. People group stars into patterns called _____ constellations _____ .

5. The solar system has thousands of _____ asteroids _____ , which are chunks of rocks or metal.

6. Our _____ solar system _____ is made up of the Sun and all the objects in orbit around it.

7. In a(n) _____ lunar eclipse _____ , Earth casts a shadow on the Moon.

8. Most _____ stars _____ are very far from Earth.

9. Galileo's _____ telescope _____ revealed objects in space that no one had seen before.

10. Once every 24 hours, Earth completes one _____ rotation _____ .

© Macmillan/McGraw-Hill

Circle the letter of the best answer for each question.

11. Which of the following is an example of apparent motion?

    A Shadows form in the presence of light.

    B Some stars appear brighter than others.

    C The Sun rises in the east.

    D The Moon orbits Earth.

12. Craters cover most of the surface of

    A Earth.

    B the Moon.

    C the Big Dipper.

    D the Sun.

13. Which of the following is an inner planet?

    A Venus

    B Jupiter

    C Saturn

    D Neptune

14. One light-year is equal to

    A the amount of time Earth travels around the Sun.

    B the distance that light travels in one year.

    C about one million miles.

    D the time it takes for one full cycle of the Moon's phases.

15. When a solar eclipse happens

    A the Moon casts a shadow on Earth.

    B there is no Moon.

    C Earth casts a shadow on the Moon.

    D there is a quarter Moon.

Answer the following questions.

16. **Observe** Why do scientists believe there is not life on any of the other planets in our solar system? Is there a planet that leads scientists to question this hypothesis?

Scientists have found no evidence of liquid water on the planets

in our solar system, except for Earth and Mars. If a planet does not

have water, it cannot sustain life. Mars, however, does have polar

ice caps and evidence of old rivers and floods. Therefore, some

scientists wonder if Mars could once have supported life.

17. **Interpret Data** Study the information in the following table.

Calendar Month	Date of Full Moon	Date of New Moon	Days in Moon Cycle	Days in Month
July	11	25	29	31
August	9	23	29	31
September	7	22	29	30
October	7	22	29	31
November	5	20	29	30

What relationship exists between the moon cycle and the number of days in a month?

The Moon completes a cycle about once a month. The full Moon

falls in the first week or two of each month shown.

_____

18. **Infer** What conclusions can be drawn about how early calendars were created? Explain.

Answers will vary but may include: Some early calendars, like the

Chinese lunar calendar, were based on the Moon cycle. People

created a system that tracks the days of the month based on the

phases of the Moon.

Answer the following questions.

19. How is the Sun different from other stars? How is it similar?

The Sun is unique because it is the closest star to Earth. Compared

to other stars, it is of average size and temperature.

20. Critical Thinking If a captain of a ship were lost at sea, how could a good understanding of the night sky help her?

Answers will vary but may include: Constellations have long been

used to help guide people at sea and on land. Before modern

technology, constellations were very important in helping people

determine their location. Today, a lost sea captain could navigate

home if she knew the position of her ship in relation to the North

Star or a familiar constellation.

21. Thinking Like a Scientist Could there be life on Neptune? Why or why not?

Neptune is far away from the Sun, making it very cold. Neptune's

cold temperatures and high winds make it an unlikely place for

living things.

Circle the letter of the best answer for each question.

1. Earth's spinning on its axis is called

   A revolution.

   B ellipse.

   C rotation.

   D orbit.

2. At what time of day are shadows the shortest?

   A in the early morning

   B at midday

   C in the late afternoon

   D just before sundown

3. Study the table about Earth below.

1 minute	=	60 seconds
1 hour	=	60 minutes
1 rotation	=	24 hours
1 revolution	=	

   Which of these belongs in the blank box?

   A 1 month

   B 1 year

   C 30 days

   D 12 hours

4. Where on Earth do the temperatures change very little from season to season?

   A the Northern Hemisphere

   B the Southern Hemisphere

   C at the North Pole

   D along the equator

Critical Thinking If you did not have a compass, how might the sunset help you find your way in the wilderness?

Possible answer: The Sun appears to set in the west. If you were lost

in the wilderness, the sunset would indicate where west is; other

directions could be determined from this information.

© Macmillan/McGraw-Hill

Name _____ Date _____

Circle the letter of the best answer for each question.

1. Which of these produces moonlight?

   A burning gases on the Moon

   B Earth's reflection off the Moon's surface

   C water reflecting on the Moon's surface

   D the Sun's reflection off the Moon's surface

2. A lunar eclipse occurs when

   A Earth's shadow falls on the Moon.

   B the Sun comes between Earth and the Moon.

   C the Moon stops spinning.

   D the Moon comes between Earth and the Sun.

3. Study the diagram below.

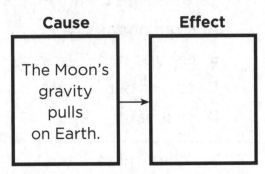

   Which of these belongs in the blank box?

   A Gravity is created on Earth.

   B The Moon is able to rotate.

   C Earth revolves around the Moon.

   D Ocean tides are created.

4. The word *phases* refers to the Moon's

   A apparent shapes.

   B gravitational pull.

   C craters.

   D distance.

Critical Thinking Why is it important for humans to protect Earth's atmosphere from pollution?

Possible answer: Earth's atmosphere helps to keep away meteoroids

that would otherwise crash into the Earth. If the atmosphere became

weakened from pollution, it might not be able to stop meteoroids from

crashing into Earth.

© Macmillan/McGraw-Hill

Circle the letter of the best answer for each question.

1. Earth travels on a(n)

   A straight path.

   B circular path.

   C square path.

   (D) oval path.

2. Which of these is the hottest planet?

   A Venus

   B Mars

   (C) Mercury

   D Saturn

3. Earth is the only planet with

   A heat.

   B a moon.

   (C) oxygen and liquid water.

   D ice and atmosphere.

4. Study the diagram below.

   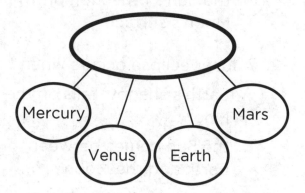

   Which of these belongs in the blank oval?

   A Moons

   B Gas Giants

   (C) Inner Planets

   D Outer Planets

Critical Thinking People who live in the city see very few stars in the sky. People who live in the country see many stars in the sky. Why do you think this is?

Possible answer: The lights in the city make the stars look very dim or

impossible to see at all. In the country it is very dark, so the stars look

very bright and are much easier to see.

Circle the letter of the best answer for each question.

1. Why does the Sun appear much larger than other stars?

   (A) It is closer than other stars.

   B It is much larger than other stars.

   C It is cooler than other stars.

   D It is warmer than other stars.

2. The Milky Way is a

   A constellation.

   B universe.

   C star.

   (D) galaxy.

3. The distance between Earth and the stars is measured in

   A miles.

   B rotations.

   (C) light-years.

   D kilometers.

4. Study the diagram below.

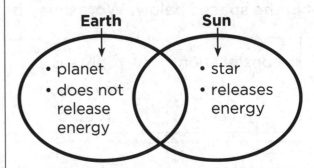

   Which of these belongs in the center section of the Venn Diagram?

   A has producers and consumers

   (B) made of layers

   C has clouds

   D is dangerous to look at directly

Critical Thinking Why do you think people long ago decided to name the constellations?

Possible answer: Before technology, people used stars to gain many

kinds of information about direction and the seasons. Since this

information was so important to them, they named the constellations

so it would be easier for them to locate the information they needed in

the sky.

© Macmillan/McGraw-Hill

The Solar System and Beyond

Write the word or words that best complete each sentence in the spaces below. Words may be used only once.

constellation	phases	solar eclipse	telescopes
craters	revolution	solar system	
gravity	rotation	stars	

1. Meteoroids created most of the Moon's ___craters___ .

2. The Sun is near the center of the ___solar system___ .

3. The Sun appears to rise and set once every time Earth completes one ___rotation___ .

4. In each complete orbit around Earth, the Moon cycles through all of its ___phases___ .

5. The Moon casts a shadow on Earth during a ___solar eclipse___ .

6. Today, scientists study space using many kinds of ___telescopes___ .

7. Each year, Earth completes one ___revolution___ around the Sun.

8. The force that pulls planets toward the Sun is ___gravity___ .

9. Draco is the name of a pattern of stars in a ___constellation___ .

10. Hot, glowing spheres of gases called ___stars___ give off energy.

© Macmillan/McGraw-Hill

Circle the letter of the best answer for each question.

11. Constellations are patterns of stars that

A depend on the observer's position on Earth.

B look the same from all points on the Earth.

C are very close together.

D cannot be seen without a telescope.

12. NASA stands for

A Natural Air and Space Association

B Natural Aeronautics and Solar Archaelogists

C National Aerial and Space Admissions

D National Aeronautics and Space Administration

13. How are solar eclipses similar to lunar eclipses?

A They both involve the Moon.

B They both cast a shadow on Earth.

C They both cast a shadow on the Moon.

D They do not involve any stars.

14. Which of these is a possible explanation for there being no water on the Moon?

A It is too cold on the Moon.

B The days are so hot that any water would evaporate.

C The Moon's gravity is too strong.

D The surface of the Moon is covered in craters.

15. The second largest planet in our solar system is <u>best</u> known for its

A moons.

B oceans and continents.

C lack of gravity.

D rings.

© Macmillan/McGraw-Hill

Answer the following questions.

16. **Interpret Data** Study the information in the following
    table.

Phase of the Moon	Tides	Position of Earth, the Moon, and Sun	Gravitational Pull of the Moon and Sun on Earth
first and third quarter	weak	Moon, Sun, and Earth form a triangle	cancel each other out
full and new Moon	strong	Earth, Sun, Moon in a line	combined

Earth's waters rise and fall each day in tides. Based on
the table above, what conclusions can be drawn about
the effect of the Moon on the tides?

When the Moon is full or new, its gravitational pull is strongest and

creates strong tides. When the Moon is in first or third quarter, its

gravitational pull is canceled out by the Sun's gravitational pull,

creating weak tides.

17. **Infer** Based on the information in the table above, can
    tides be predicted and planned for in advance?

    Yes, the phases of the Moon can be tracked and can help to

    predict weak and strong tides.

18. **Observe** There is evidence that water once flowed on
    Mars. Though temperatures are too low on Mars today
    for liquid water, what does this suggest about the
    planet's past?

    Answers will vary but may include: Evidence of water on Mars

    suggests that the planet may have once been able to sustain life.

    Scientists' research of the ice on Mars may reveal if this was true.

_____

Answer the following questions.

19. How are meteoroids, meteors, and meteorites different?

Meteoroids are pieces of asteroids that enter Earth's atmosphere;

small meteoroids burn up in Earth's atmosphere and leave streaks

of light called meteors. Any of these fragments from space that

strike Earth's surface are called meteorites.

20. Critical Thinking Do constellations appear the same in the Northern and Southern Hemispheres? Why or why not?

Constellations depend on the observer's position on Earth and will

appear to be different if someone is looking at the night sky from

the Northern Hemisphere or from the Southern Hemisphere. A few

constellations appear in both hemispheres.

21. Thinking Like a Scientist Identify the four rocky planets in our solar system. What do they have in common?

Mercury, Venus, Earth, and Mars are the four rocky planets. Rocky

planets are mostly made up of rock and have solid cores made of

iron.

The Solar System and Beyond

Materials

- paper
- markers
- computer

Make a Star Chart

Objective: Students will do research on the Internet about the stars. They will gather information about the size, color, life span, and temperature of stars in the galaxies. They will create a colorful and informative chart with colored pencils or markers that reflects this information.

Scoring Rubric

4 points Student conducts research on the Internet, finds the information needed to create a star chart, and creates an informative and colorful chart about stars. Chart includes color, temperature, size, and life span of the stars researched. Answers to questions in Analyze the Results are accurate and correct.

3 points Student conducts research on the Internet, finds some of the information needed to create a star chart, and creates a chart about stars. Chart includes color, temperature, size, and life span of the stars researched. Answers to questions in Analyze the Results are mostly accurate and correct.

2 points Student tries to conduct research on the Internet, but finds little of the information needed to create a star chart and the chart is incomplete. Answers to questions in Analyze the Results are mostly incorrect.

1 point Student attempts the assignment but the chart and the answers to the questions are mostly incorrect.

Make a Star Chart

Conduct research on the Internet about stars. Look for Web sites that will help you find the average age, size, temperatures, and life spans of stars. Gather the information and create a colorful and informative chart. The chart can be a line graph, a bar graph, a labeled drawing, a pie chart, or a table. Make sure to label the chart so other students can interpret the information.

> **Internet Tips to Stay Smart and Safe**
> ✔ DO visit Web sites that will help you with your project.
> ✔ DO ask your teacher for help if you get lost.
> ✘ DO NOT talk to strangers on the Internet.

Analyze the Results

1. What Web sites did you visit to do research about stars?

Answers will vary. Check that the Web sites visited are up-to-date

and reliable.

_____

_____

2. What did you find out about stars? Write a brief summary of the results of your research here.

Answers will vary. Make sure the results described here coincide

with the chart the student created. Also, double-check that the

student's chart is not a reproduction of one found on the sites

visited.

Properties of Matter

Write the word that best completes each sentence in the spaces below. Words may be used only once.

area	element	matter	volume
atom	length	metals	
density	mass	properties	

1. Anything that has mass and takes up space is _____matter_____ .

2. A(n) _____element_____ is a substance that is made up of only one type of matter.

3. An object's _____length_____ is the distance from one end to the other end.

4. The amount of matter in an object is its _____mass_____ .

5. The number of unit squares that cover a surface is called the _____area_____ .

6. Materials that can be bent or hammered into shape and conduct heat and electricity are _____metals_____ .

7. How much space an object takes up is its _____volume_____ .

8. The amount of mass in a unit of volume is its _____density_____ .

9. Smell, color, texture, hardness, and shape are all examples of _____properties_____ .

10. A(n) _____atom_____ is the smallest particle of an element that still has all the properties of that element.

© Macmillan/McGraw-Hill

Circle the letter of the best answer for each question.

11. A gas has

   A mass, volume, and state.

   B volume only.

   C mass only.

   D mass and definite shape.

12. How is the density of an object calculated?

   A divide its state by its area

   B multiply its mass by its matter

   C divide its volume by its state

   D divide its mass by its volume

13. Matter can be defined as

   A millions of atoms of the same element.

   B made up of hydrogen and carbon.

   C anything that has mass and takes up space.

   D only visible materials.

14. Buoyancy is the property of matter that can be defined as

   A being able to dissolve into liquids.

   B the ability to conduct heat or electricity.

   C the upward force of a liquid or gas.

   D whether or not light can pass through it.

15. What does weight measure?

   A the pull of gravity between an object and a planet

   B the force of the pull between objects

   C the level of an object's buoyancy

   D the density of an object's volume

Answer the following questions.

16. **Infer** What might happen to a liquid with a freezing point of 32 degrees if its temperature dropped below 30 degrees?

    The liquid would freeze and turn into a solid.

    _____

    _____

17. **Measure** What is calculated by multiplying the length of a rectangle by its width and height?

    volume

    _____

18. **Classify** Describe the physical properties of a liquid and a solid. How are they alike? How are they different?

    Answers will vary but may include the following: A solid has a

    definite shape and takes up a definite amount of space. A liquid

    takes up a definite amount of space but does not have a definite

    shape.

Answer the following questions.

19. **Critical Thinking** Below is a section of the periodic table of elements. How are the elements grouped? Explain your answer with an example.

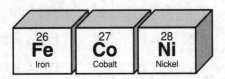

Elements are grouped by their properties. For example, iron,

cobalt, and nickel are metals and are the most magnetic elements

in the periodic table.

20. **Thinking Like a Scientist** Flour is a substance used to bake breads and cakes. Describe an experiment that would scientifically prove that flour should be classified as a solid and not as a liquid or gas.

Answers will vary but may include the following: The theory

that flour is a solid could be tested by determining if flour has a

definite shape and takes up a definite amount of space. Though

flour conforms to the shape of a container, as a liquid does, this

experiment will test to see what happens if the container is placed

upside-down on a tray. When the container is removed slowly, the

flour retains the shape of the container. A liquid would not.

Circle the letter of the best answer for each question.

1. Light is not matter because

   A it is buoyant.

   B it travels very fast.

   (C) it does not take up space.

   D it is the combination of many colors.

2. If an object is buoyant, then it will

   (A) float.

   B sink.

   C attract metal.

   D lose its shape.

3. Which of these contains the most tightly packed matter particles?

   A tea

   B carbon dioxide

   C a milkshake

   (D) cardboard

4. Study the chart below.

   Properties of Matter

Solid	Liquid	Gas
wood	milk	helium
sand	vegetable oil	

   Which of these belongs in the "Gas" category?

   A salt

   (B) oxygen

   C water

   D soil

Critical Thinking Do you think it is a good idea to reuse matter? Explain your answer.

Possible answer: It is a good idea to reuse matter. If we reuse matter it

reduces the amount of trash that ends up in landfills. Reusing matter is

one way we can help to save the environment.

© Macmillan/McGraw-Hill

Circle the letter of the best answer for each question.

1. Why do objects weigh less on the Moon than they do on Earth?

   A Objects do not have as much mass on the Moon.

   B The area of an object decreases on the Moon.

   C There is a different scale of measurement used on the Moon.

   D The pull of gravity is not as strong on the Moon.

2. The metric unit for weight is called a

   A ton.

   B kilometer.

   C newton.

   D liter.

3. Which of these is an example of volume?

   A the number of cubes that fit inside a box

   B the number of squares that cover a surface

   C the number of units that fit from one end of a line to the other

   D number of units that fit across a shape from one side to the other

4. Study the problem below.

   Mass ÷ _____ = Density

   Which of the following belongs in the blank space?

   A Buoyancy

   B Volume

   C Area

   D Weight

Critical Thinking A two-ounce, solid steel marble is placed in a bathtub full of water, and it sinks to the bottom. An empty, two-pound steel bowl is placed in the bathtub and it floats. Why?

The air inside the bowl makes it more buoyant than the steel ball. The

density of the bathtub water is greater than that of the empty bowl,

so the bowl floats. But the water's density is less than that of the solid

marble, so the marble sinks.

Circle the letter of the best answer for each question.

1. The periodic table of elements organizes elements into columns and rows called

   A matter and atoms.

   B solids and liquids.

   C knowns and unknowns.

   (D) groups and periods.

2. A metalloid has

   (A) some of the properties of a metal, but not all.

   B none of the properties of a metal.

   C all of the properties of a metal.

   D no clear position on the periodic table.

3. Study the chart below.

Metals	Nonmetals
aluminum	nitrogen
copper	carbon dioxide
iron	

   Which of these belongs in the empty box?

   A gold

   (B) neon

   C nickel

   D zinc

4. Which of these is <u>not</u> an element?

   A oxygen

   B carbon

   (C) water

   D nitrogen

Critical Thinking How do you think the modern day periodic table of elements compares to the original table that Dmitry Mendeleev created 150 years ago? How does it look different?

Today's table is probably bigger because Mendeleev did not know all the

elements that we know today. Today's table probably has more groups

than the old table did. Some of the elements may have moved to different

places on the table as scientists learned more about them.

Properties of Matter

Write the word that best completes each sentence in the spaces below. Words may be used only once.

area	element	matter	volume
atoms	length	metals	
density	mass	property	

1. A large bottle of water has more _____volume_____ than a small glass of water.

2. A(n) _____element_____ is a kind of matter that cannot be broken down into anything simpler.

3. To find an object's _____area_____ , one may draw and count one-inch squares across its surface.

4. Each element is made up of tiny units called _____atoms_____ .

5. People use _____length_____ to measure things, such as height.

6. Heat and electricity can pass through _____metals_____ .

7. How tightly matter is packed together can be described as _____density_____ .

8. A large rock has more _____mass_____ than a small rock.

9. Something that has mass and occupies space can be described as _____matter_____ .

10. A(n) _____property_____ is any characteristic of matter that can be observed in some way.

© Macmillan/McGraw-Hill

Circle the letter of the best answer for each question.

11. The number of units that fit from one end of an object to the other determine its

    A density.

    B area.

    C state.

    (D) length.

12. Which of these <u>best</u> describes the physical properties of matter?

    A All matter has density and volume.

    B Most matter has mass, state, and area.

    (C) All matter has mass, volume, and state.

    D Most matter has volume and form.

13. The number of cubes that fit inside an object determines its

    A mass.

    B area.

    C density.

    (D) volume.

14. Particles that can be classified by their properties make up

    (A) elements.

    B matter.

    C atoms.

    D mass.

15. Which of the following is the <u>most</u> dense?

    A water

    (B) sand

    C cork

    D oil

Answer the following questions.

16. **Infer** What might happen to a solid with a melting point of 100 degrees if it were dropped in boiling water?

The solid would melt and become a liquid.

17. **Measure** How is the area of a rectangle calculated?

Area is calculated by multiplying the rectangle's length by its

width.

18. **Classify** Describe the physical properties of a gas. How does a gas compare to a liquid?

A gas does not have a definite shape and does not take up a

definite amount of space. A liquid takes up a definite amount of

space but does not have a definite shape.

Name _____ Date _____

Answer the following questions.

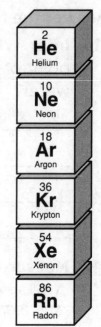

19. **Critical Thinking** Group 18 on the Periodic Table of Elements contains the Noble gases. The word *noble* means two things: being a member of a high class, or royalty; and being chemically inactive. What do Noble gases have in common, and could their group name use both definitions of the word?

Answers will vary but may include the following:

The elements in Group 18 are all gases that rarely

participate in chemical reactions. They are called

"noble" because they are chemically inactive, but

they could also have that name because they belong

to an elite class of elements.

20. **Thinking Like a Scientist** Flour is a solid powder used to make cakes and breads. It is made up of very small, solid pieces. Hypothesize about what would happen to baking flour if it were mixed with an equal amount of water. How can this new substance be classified?

Answers will vary but may include the following: Having

determined that baking flour is a solid and water is a liquid, I

hypothesize that if the two items were mixed together in equal

parts, they would form a solid because the powder of the flour

would absorb the liquid water and create a paste which would

have a definite shape and a definite volume.

Changing States

Objective: Students will experiment with the various states of water. Students will begin with a solid block of ice. They will then observe as the ice melts into water in its liquid state. Finally, students will evaporate the water, observing that it turns into a gas.

Materials

- small block of ice
- metal tray
- pen
- paper

Scoring Rubric

4 points   Student handles the block of ice carefully and as directed in the objective. Student records observations on a piece of paper as the block of ice melts into water and then evaporates. Student clearly answers the questions in Analyze the Results. Answers to all of the questions are correct.

3 points   Student handles the block of ice carefully and as directed in the objective. Student records some observations on a piece of paper. Student's answers to questions are partially correct.

2 points   Student handles the block of ice carelessly. Student records a few observations on a piece of paper. Answers to the questions are mostly incorrect.

1 point   Student attempts the assignment but records no observations and the answers to the questions are incorrect.

Changing States

Experiment with Water

Use the materials supplied by your teacher to investigate the various states of water. Write down your observations on a piece of paper.

- First, investigate the block of ice. Feel how cold it is, note how solid it is. It has a definite shape and a definite volume. Watch it melt as you hold it in your hands or as it sits in the tray on your desk.

- Next, allow the block of ice to melt into water as it sits at room temperature in the tray on your desk. Feel the water. Notice that it has no definite shape, but that it has a definite volume.

- Finally, label your tray with your name and place it on top of a radiator or in a sunny window. Leave the tray there overnight. Check the tray in the morning. The water will have disappeared because it has turned into a gas. The gas has no definite shape and no definite volume.

Analyze the Results

1. Describe what happened to the block of ice. In which states did you observe the water?

   The water started out in a solid state, then slowly melted into its

   liquid state, and finally evaporated into its gas state.

2. Could the water have gone through each of the states in a different order? If so, how?

   It would be possible to collect water as gas, condense it into its

   liquid state, and then freeze it into its solid state.

   _____

   _____

Name _____ Date _____

Matter and its Changes

**Write the word that best completes each sentence
in the spaces below. Words may be used only once.**

acids	chemical change	evaporation	tarnish
alloy	compounds	filter	
bases	distillation	physical change	

1. Chopping wood is an example of a(n) __physical change__ .

2. A(n) __alloy__ is a mixture of two or more
   metals, or a mixture of metals and nonmetals.

3. A process called __distillation__ can separate salt
   and water from salt water.

4. The change of a liquid into a gas is called __evaporation__ .

5. Weak __acids__ are found in many foods,
   such as lemons and oranges.

6. A piece of mesh or a screen that holds back large

   pieces in a mixture is a(n) __filter__ .

7. When some metals react chemically with sulfur in the

   air, __tarnish__ forms.

8. Substances that taste bitter and have a slippery,

   soapy feel are __bases__ .

9. Burning a match is an example of a(n) __chemical change__ .

10. Two or more elements that are chemically combined

    are __compounds__ .

Circle the letter of the best answer for each question.

11. Which of the following is <u>not</u> an example of a physical change?

    A clay is molded into a new shape

    B paper is cut into pieces

    C water evaporates into gas

    (D) an iron nail rusts

12. Which statement is false?

    (A) A solution can never be separated.

    B Materials in a mixture do not blend completely.

    C An alloy always contains at least one metal.

    D The air people breathe is a solution of gasses.

13. Which of the following takes place when a substance undergoes a chemical change?

    A a change of state

    B a change of shape

    C a change of size

    (D) a change of matter

14. Which of the following is a true statement about compounds?

    A Two or more elements can physically combine to create a unique element.

    B Compounds are made up of elements that exist in the same state as the resulting compound.

    (C) When elements join to form a compound they lose their original properties.

    D Compounds are created only in laboratories and never appear in nature.

15. Settling and filtration are examples of

    A ways to separate a solution.

    B methods used to turn saltwater into drinking water.

    (C) ways to separate a mixture.

    D methods used to return an alloy to its original metals.

© Macmillan/McGraw-Hill

Answer the following questions.

16. **Observe** How are physical and chemical changes alike? How are they different? Fill in the Venn diagram below.

Physical Changes **Both** **Chemical Changes**

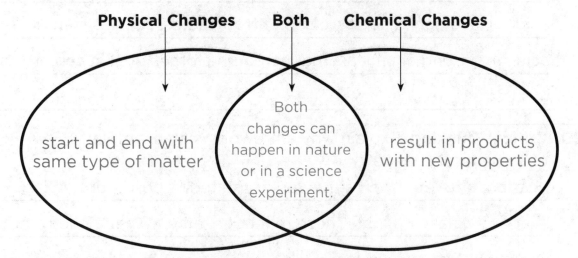

start and end with same type of matter

Both changes can happen in nature or in a science experiment.

result in products with new properties

17. **Interpret Data** A student mixes salt with warm water in a large container. What is salt mixed with water an example of?

Salt water is a solution in which the salt and the water combine

completely when stirred together.

18. Explain what a compound is and give an example. Include a description of the original parts of your example and the resulting compound.

Answers will vary but may include: A compound is a substance

made up of two or more elements that have chemically combined.

Water is a compound. It is made up of hydrogen and oxygen.

When they combine, they form liquid water.

Answer the following questions.

19. How does lemon juice compare to sulfuric acid?

Lemon juice and sulfuric acid are both examples of acids. Weak

acids, like lemon juice, taste sour. Strong acids, such as sulfuric

acid, are strong chemicals that can burn clothes or skin and are

unsafe to taste.

20. Critical Thinking Explain why a firework is an example of a chemical change?

Possible answer: When a firework is ignited, it burns, producing

heat and a flash of light. These reactions tell you that a chemical

change has occurred.

21. Thinking Like a Scientist A scientist is conducting an experiment with sand. Some iron filings accidentally fell into the sand. What could the scientist use to separate out the two substances?

Possible answer: The scientist could use a magnet to attract the

iron filings and remove them from the sand.

Circle the letter of the best answer for each question.

1. A change of state does not affect an object's

(**A**) mass.

B volume.

C shape.

D texture.

2. When a liquid is cooled, its particles become

A farther apart.

B very fast moving.

C melted.

(**D**) closer together.

3. When water is boiled, it changes from

(**A**) a liquid to a gas.

B a solid to a liquid.

C a gas to a solid.

D a solid to a gas.

4. Study the diagram below.

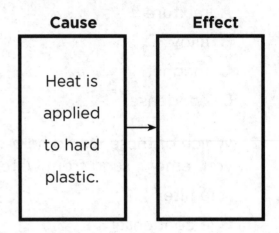

Which of these <u>best</u> fits in the blank box?

A Plastic remains solid on the surface but turns liquid on the inside.

B Plastic turns from liquid to solid.

(**C**) Plastic turns from solid to liquid.

D Plastic freezes.

Critical Thinking Imagine that there was no rain for two weeks and that it was very sunny. How would this affect the water level in the local pond? Explain your answer.

Possible answer: The water level would go down because heat from

the Sun would evaporate the water. When water evaporates, it

changes into a gas and rises up into the air.

_____

Name _____ Date _____

Circle the letter of the best answer for each question.

1. Steel is an example of a(n)

   A mixture.

   (B) alloy.

   C magnet.

   D condenser.

2. Which of these would help you remove sand from water?

   (A) a filter

   B a condenser

   C a magnet

   D a mixture

3. Which of these is a solution?

   A several different kinds of nuts in a jar

   B a box full of pens and pencils

   (C) chocolate syrup in a glass of milk

   D a denim jacket with a leather collar

4. Study the diagram below.

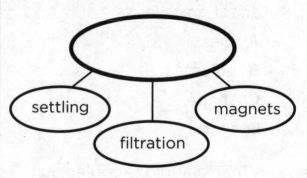

   Which title belongs in the empty oval?

   A How to Separate a Solution

   B Parts of an Alloy

   (C) How to Separate a Mixture

   D Parts of a Solution

Critical Thinking In what ways are evaporation and distillation different? In what ways are they similar?

Possible answer: Evaporation and distillation both change a liquid into

a gas. They are different because evaporation is a way of collecting

solids, and distillation is a way of collecting liquids.

Circle the letter of the best answer for each question.

1. If two elements are chemically combined, they

   A always create a solid.

   (B) lose their original properties.

   C always create a liquid.

   D keep their original properties.

2. Which of these <u>best</u> describes a compound?

   (A) two or more elements chemically combined

   B elements that rust when exposed to oxygen

   C two elements that cannot be combined

   D a group of elements that create heat when combined

3. Which of these must be used in order to separate chemicals in a compound?

   A magnets

   B filtration

   C settling

   (D) a chemical reaction

4. Which of the following is not a compound?

   A water

   B rust

   (C) iron

   D salt

Critical Thinking If your tongue feels like it is burning from eating a spicy pepper, drinking milk will make the burning feeling stop. Why do you think this is?

Possible answer: Pepper juice is an acid and milk is a base. Drinking

milk neutralizes the pepper acid so it doesn't burn.

Matter and its Changes

Write the word that best completes each sentence in the spaces below. Words may be used only once.

acids	compound	mixture	tarnish
alloy	evaporation	rust	
bases	filter	solution	

1. A chemical reaction that causes silver to turn black when exposed to the air for too long is _____tarnish_____ .

2. Strong _____acids_____ are dangerous chemicals that can burn skin and dissolve some metals.

3. Bronze is a(n) _____alloy_____ made of copper and tin.

4. The result of a chemical reaction between iron and oxygen is called _____rust_____ .

5. A mixture in which one or more substances blends completely in another kind of matter is called a(n) _____solution_____ .

6. Household items such as soap, detergent, and baking soda are _____bases_____ .

7. A(n) _____filter_____ can be used to separate the parts of some mixtures.

8. Water is a(n) _____compound_____ made up of hydrogen and oxygen.

9. When the Sun turns water into a gas, it is called _____evaporation_____ .

10. Two types of matter that are combined, but keep their original chemical properties, is a(n) _____mixture_____ .

Circle the letter of the best answer for each question.

11. Which statement is true?

    A Tarnish forms when silver reacts with hydrogen.

    (B) The release of energy is a sign that a chemical change occurred.

    C Evaporation collects the solids in a solution.

    D Weak acids are not found in foods.

12. Which of the following is an example of a chemical change?

    A crushing a tin can

    (B) cooking eggs

    C ice melting into water

    D stretching a rubber band

13. Which of the following methods separates a liquid from a solution?

    (A) distillation

    B condensation

    C filtration

    D evaporation

14. What are two processes that can change a liquid into a gas?

    (A) boiling and evaporation

    B melting and boiling

    C cooling and freezing

    D evaporation and freezing

15. Which of the following takes place when a substance undergoes a physical change?

    A a change in color

    B a change in direction

    C a change in matter

    (D) a change in size

Answer the following questions.

16. **Observe** Think of a physical and chemical change. Fill in the chart below with the cause and effect of each change. Explain your answers.

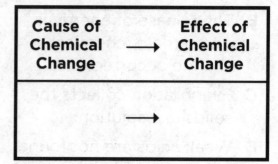

Answers will vary but may include the following. Physical Change —

Cause: clay is molded into the shape of a bowl. Effect: matter, mass,

and volume of the clay stay the same. Chemical Change — iron nail

is left outside. Effect: iron reacts with oxygen in air and changes

into rust.

17. **Interpret Data** A student fills a salad bowl with carrots, tomatoes, peppers, and lettuce. Is this a mixture, a solution, or a compound? How do you know?

A salad is an example of a mixture. The parts do not combine

completely, and they can be easily separated.

Answer the following questions.

18. The elements sodium and chlorine combine chemically to produce table salt. Is table salt a compound, a mixture, or a solution? How is table salt different from the elements from which it is made?

Table salt is a compound. It is edible and can be categorized as a

nonmetal. Sodium is a metal, and chlorine is a poisonous green gas.

19. Fill in the chart at right with an example of an acid and three of its characteristics.

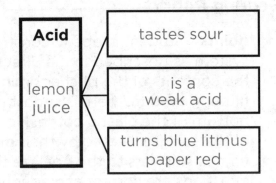

Acid

lemon juice

tastes sour

is a weak acid

turns blue litmus paper red

20. **Critical Thinking** How is baking a cake an example of a chemical change?

When a cake is baked, the heat from the oven produces a

chemical change that changes the cake batter's taste, texture, and

color. Before the batter is cooked, it is wet and sticky. After it is

cooked, it is warm and fluffy.

21. **Thinking Like a Scientist** Vinegar and baking soda have been combined. There are bubbles rising to the surface of the mixture. What conclusions can be drawn about this experiment? Was there a chemical reaction?

There was a chemical reaction between the vinegar and baking

soda. A gas was released, and a solution was formed.

© Macmillan/McGraw-Hill

Filters and Mixtures

Objective: Students will make their own filter system. They will choose a filtering material and pour the contents of the pitcher into the funnel. When they have let it drain they will examine the filtered water and decide whether the material they have chosen separated the soil from the water.

Materials

- soil
- water
- plastic pitcher
- funnel
- clear plastic soft drink bottle
- cotton
- pebbles
- paper towels

Scoring Rubric

4 points   Student properly constructs the filter system as described in the directions. When the contents of the pitcher drain through their filtering system, there is almost no soil in the bottom of the plastic bottle. Most, if not all, of the soil is collected by their material in the funnel. Answers to the Analyze the Results questions are correct and adequately represent the experiment.

3 points   Student properly constructs the filter system as described in the directions. When the contents of the pitcher drain through their filtering material, there is some soil in the bottom of the plastic bottle. Most of the soil is collected by their material in the funnel. Answers to the Analyze the Results questions are mostly correct and represent the experiment.

2 points   Student does not properly construct the filter system. When the contents of the pitcher drain through their filtering material, there is a large amount of soil in the bottom of the plastic bottle. A small amount of soil is collected by their material in the funnel. Answers to the Analyze the Results questions are mostly incorrect and do not adequately represent the experiment.

1 point   Student attempts the model but cannot construct the filter system and the answers to the Analyze the Results questions are incomplete.

Filters and Mixtures

Make a Model

From the materials provided by your teacher, you will construct a filter system. Choose the best material to put into the funnel to act as a filter. Place the funnel into the clear soft drink bottle. Now, fill the plastic pitcher with water and stir the soil in. Next, slowly pour the contents of the pitcher into the funnel and let it drain through. Examine the water that you have filtered.

Analyze the Results

1. Which material did you choose? Did the material filter the water contents? Why or why not?

   Answers will vary depending on the material the student selected.

   The cotton and paper towels will most likely do the best job of

   filtering the soil from the water. Students should also mention the

   size of the holes in the material they used as a filter. The smaller

   the holes, the better the filtering system.

2. Is the soil and water combination a mixture or a solution?

   Answers will vary but may include: The soil and water is a mixture

   because it can be separated into parts. Students may also mention

   that it is possible that minerals or salts in the soil could have been

   dissolved into a solution.

Forces

Write the word or words that best completes each sentence in the spaces below. Words may be used only once.

acceleration	kinetic energy	unbalanced forces
balanced forces	newtons	velocity
compound machine	potential energy	
gravity	speed	

1. An object's speed and direction of motion is its
   _____velocity_____ .

2. Forces that are equal in size and are opposite in
   direction are ____balanced forces____ .

3. The distance an object moves in a certain amount of
   time is _____speed_____ .

4. Force is measured in units called _____newtons_____ .

5. Stored energy is called ____potential energy____ .

6. The force that acts over a distance and pulls all
   objects together is _____gravity_____ .

7. The energy of motion is ____kinetic energy____ .

8. A change of speed or direction of an object's motion
   is called _____acceleration_____ .

9. Forces not equal in size, or opposite in direction are
   _____unbalanced forces_____ .

10. Simple machines can be combined to make a(n)
    _____compound machine_____ .

Circle the letter of the best answer for each question.

11. How can you determine the speed of a moving train?

(A) divide the distance traveled by the time spent moving

B divide the time spent moving by the direction traveled

C multiply the distance traveled by the time spent moving

D multiply the velocity by the distance traveled

12. Friction can <u>best</u> be described as

A a force that helps move objects.

B a force found between a car and its tires.

C a force that is not affected by gravity.

(D) a force that works against motion.

13. How do scientists define the term *work*?

A energy spent to complete a task

(B) a force used to move an object a certain distance

C energy stored and later released

D a measure of how much energy a machine produces

14. Which of the following is an example of a compound machine?

A lever

B pulley

(C) scissors

D wedge

15. Which of the following statements is false?

A Energy is the ability to do work.

B Energy can be either potential or kinetic.

C Energy can be transferred from one object to another.

(D) Energy cannot change from potential to kinetic.

Answer the following questions.

16. **Interpret Data** A speed skater is racing around a track. Read the steps of her race and fill in the chart with a description of what is happening. The first one is done already. Choose your answers from the words in the box.

acceleration	friction	inertia

She pushes her feet against the ice to gain momentum.	**force**
She moves to the left as the track curves.	acceleration
After she crosses the finish line she stops pushing her feet, but she still moves forward.	inertia
After a while she leans back onto the rough section of her skate's blades and stops moving.	friction

17. **Infer** List two or more simple machines from daily life. How would the world be different without these simple machines?

Possible answer: A doorstop keeps the classroom door open. This

is an example of a wedge. Screws hold the desks together in the

classroom. A screw is an inclined plane twisted in a spiral. Without

these tools people would have to exert much more energy to do

simple tasks.

Answer the following questions.

18. What happens to the energy in a bicycle as it moves slowly up a hill? What happens to it as it starts to move down the other side of the hill?

As the bicycle moves up the hill, it stores energy. This is called

potential energy. As the bicycle starts to move down the other

side of the hill, the energy is transferred into kinetic energy, and

the bicycle moves very quickly.

19. Critical Thinking Energy is constantly changed from one form to another. Give an example of a household object that transforms energy.

Answers will vary but may include: Objects such as light bulbs

transform electrical energy into heat and light energy.

20. Thinking Like a Scientist Fossil fuels play a major role in the development of electrical energy. Why is electrical energy so important?

Answers will vary but may include: Electrical energy is a necessity

to the daily life of most people on Earth. Without it, it would be

much harder or even impossible to keep food cold, read at night,

listen to music, or do research on a computer.

Name _____ Date _____

Circle the letter of the best answer for each question.

1. Velocity refers to

    A the distance and time of motion.

    B downward force.

    Ⓒ the speed and direction of motion.

    D the amount of force.

2. Frame of reference refers to

    Ⓐ position.

    B force.

    C speed.

    D friction.

3. What information is needed in order to calculate speed?

    A the distance traveled

    B time and the amount of force exerted

    C the mass of the object

    Ⓓ the distance traveled and time

4. Study the diagram below.

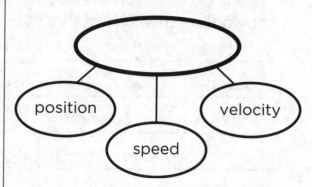

Which of these belongs in the blank oval?

    A Types of Force

    B Frame of Reference

    C Phases of Inertia

    Ⓓ Ways to Describe Motion

Critical Thinking If you rolled a ball across the ground on the Moon, how would it compare to a ball rolled on Earth?

Possible answer: The ball on the Moon would roll farther than the ball

on Earth because there is less gravity on the Moon. The ball would also

bounce much higher on the Moon, because the pull of gravity is weak.

Name _____ Date _____

Circle the letter of the best answer for each question.

1. A newton is a unit that measures

   A force.

   B weight.

   C distance.

   D speed.

2. When two children of equal weight sit at opposite ends of a seesaw, they create

   A gravity.

   B unbalanced forces.

   C motion.

   D balanced forces.

3. All of these affect inertia <u>except</u>

   A size.

   B mass.

   C color.

   D shape.

4. Study the diagram below.

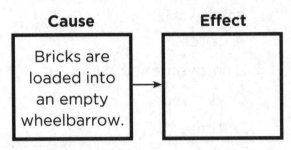

Cause		Effect
Bricks are loaded into an empty wheelbarrow.	→	

   Which of these belongs in the blank box?

   A Less force is required to move the wheelbarrow.

   B The amount of friction decreases.

   C More force is required to move the wheelbarrow.

   D The wheelbarrow's speed increases.

Critical Thinking A chef coats cooked spaghetti with oil before putting sauce on the spaghetti. What force is the chef trying to control? Explain your answer.

Possible answer: The chef is trying to control friction. The oil is slippery,

so it decreases the friction between the spaghetti noodles. If oil were

not used, the spaghetti noodles would stick together.

© Macmillan/McGraw-Hill

Circle the letter of the best answer for each question.

1. Work involves all of these <u>except</u>

   A force.

   B movement.

   C distance.

   (D) time.

2. Which of these is <u>not</u> true about potential energy?

   A It can change to kinetic energy.

   (B) It always involves gravity.

   C It is energy that is gained prior to work.

   D It sometimes involves gravity.

3. Energy that comes from the tiniest particles of matter is called

   A chemical energy.

   B thermal energy.

   (C) nuclear energy.

   D light energy.

4. Study the diagram below.

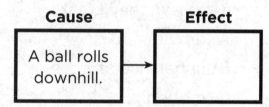

Cause	Effect
A ball rolls downhill. →	

   Which of these belongs in the blank box?

   (A) Kinetic energy is created.

   B Potential energy is created.

   C Thermal energy is created.

   D Energy is transferred from ball to ground.

Critical Thinking Are you doing work when you blow a bubble with chewing gum? Explain your answer.

Possible answer: Yes, work is the use of force to move an object. When you blow a bubble, your breath is moving and stretching the chewing gum.

Circle the letter of the best answer for each question.

1. A playground seesaw is a
   type of

   A pulley.

   (B) lever.

   C wheel and axle.

   D compound machine.

2. Which of these describes
   effort force?

   A an object being moved by
   a lever

   B the fixed point of a lever

   (C) the force used to do work

   D the measure of a machine's
   efficiency

3. Which of these is an example
   of a wheel and axle?

   (A) a ceiling fan

   B a hammer

   C a pair of scissors

   D a shovel

4. Which of these reduces the
   efficiency of a machine?

   A distance

   B work

   C speed

   (D) friction

Critical Thinking Imagine that you needed to roll a
large boulder several feet. What simple machine would
you use to perform this task? Explain your answer.

Possible answer: A lever would be a good simple machine to move

a rock. One end of the lever could be wedged under the rock, which

would help to start it roll when effort force was applied to the other

end of the lever.

Forces

Write the word or words that best complete each sentence in the spaces below. Words may be used only once.

balanced forces	inertia	unbalanced forces
compound machines	kinetic energy	velocity
friction	potential energy	
inclined plane	speed	

1. Forces called ___unbalanced forces___ cause a change in motion.

2. How fast an object is moving in a specific direction describes its ___velocity___ .

3. Most machines, like scissors, are ___compound machines___ .

4. Forces called ___balanced forces___ do not cause a change in motion because they cancel each other out.

5. To find an object's ___speed___ , divide the distance traveled by the time spent moving.

6. The future ability to do work is ___potential energy___ .

7. A rock will not move unless a force acts on it because of its ___inertia___ .

8. There is little ___friction___ between ice and the blades on ice skates.

9. A(n) ___inclined plane___ is a straight, slanted surface.

10. If an object is moving, it has ___kinetic energy___ .

Circle the letter of the best answer for each question.

11. Which of the following <u>best</u> describes energy?

    A the ability to remain in motion

    (B) the ability to do work

    C the ability to pull objects down

    D the ability to change states

12. Which of the following is an example of a simple machine?

    A bicycle

    B scissors

    (C) ramp

    D can opener

13. The tendency of an object to stay in motion or at rest is called

    A velocity.

    B acceleration.

    C speed.

    (D) inertia.

14. The force used to do work is called

    A friction.

    B unbalanced force.

    (C) effort force.

    D negative force.

15. When a roller coaster moves down a ramp

    A kinetic energy changes to potential energy.

    B electrical energy changes to light energy.

    (C) potential energy changes to kinetic energy.

    D gravity is not involved.

Answer the following questions.

16. **Interpret Data** Read the data below about a roller coaster. Fill in the chart with a description of what kind of energy is being used and whether the roller coaster is doing work. The first one is done already. Choose your answers from the words in the box.

kinetic energy	does work
potential energy	does not do work

The roller coaster leaves the gate.	kinetic energy	does work
The roller coaster goes up the ramp.	kinetic energy	does work
The roller coaster stops at the top of the ramp.	potential energy	does not do work
The roller coaster races down the ramp.	kinetic energy	does work

17. **Infer** In some parts of the world there are wild monkeys that use tools. Imagine what types of tools these animals might create, and explain how they are similar to objects we use in our everyday lives.

Answers will vary but may include: Wild monkeys probably create

tools very much like the simple machines we use everyday; they

might use a piece of wood as a lever.

_____

Answer the following questions.

18. Explain why light energy is important to all living things on Earth.

The Sun is the major source of light energy on Earth. Plants use

light energy from the Sun to make food. During photosynthesis,

light energy is converted to chemical energy as food for plants.

Animals rely on plants as an important food source.

19. Critical Thinking Why does a rough surface cause greater friction to an object moving across it than a smooth surface? Do any smooth surfaces cause more friction than some rough surfaces?

Possible answer: Most rough surfaces cause greater friction

because there is more surface area in the bumps and cracks for

the object to come into contact with. The bumps in the rough

surface grab hold of the object so it doesn't slide as easily. Rubber

is an example of a smooth surface that causes a lot of friction.

20. Thinking Like a Scientist What problems are produced by burning fossil fuels? What alternative resources could we use to produce electricity?

Fossil fuels are nonrenewable resources and they cause pollution.

We can produce electricity using renewable resources such as the

Sun, wind, and water.

Forces

Materials

- set of building blocks
- string
- popsicle sticks
- glue
- cardboard
- paper
- scissors

Make a Compound Machine

Objective: Students will build their own compound machine out of two or more simple machines. Students should first build two simple machines such as a lever, a pulley, a wheel and axle, an inclined plane, or a wedge. Students must then combine the two machines to create a compound machine. Students must be able to show how the compound machine works and explain it to the class.

Scoring Rubric

4 points Student builds two or more simple machines and correctly puts them together to create a working compound machine. Student is able to explain to the class how the compound machine works. The compound machine does the work the student intended. Answers to questions in Analyze the Results are correct.

3 points Student builds two or more simple machines and puts them together to create a compound machine. Student is able to explain to the class how the compound machine works. The compound machine mostly does the work the student intended. Answers to questions in Analyze the Results are mostly correct.

2 points Student builds two simple machines and puts them together more or less correctly. Student is unable to explain to the class how the compound machine works. The compound machine does not do the work the student intended. Answers to questions in Analyze the Results are mostly incorrect.

1 point Student attempts the assignment but the compound machine does not work, or is instead a simple machine. Answers to the questions are incorrect.

Make a Compound Machine

Make a Model

Choose from the materials provided by your teacher to make two or more simple machines that you will then combine to build a compound machine. Remember that your machine has to be able to do work. Think about the characteristics of the machine before you start to build your own, and what work your machine will perform. What simple machines would best fit together to do this work?

Analyze the Results

1. Describe your compound machine. Make sure to explain the uses for each of its parts.

   Answers will vary. Students should explain all of the simple

   machines that were used to build the compound machine.

   _____

   _____

   _____

   _____

2. What work does your compound machine perform? What does each simple machine contribute to the compound machine?

   Answers will vary but must include details about the role of each

   simple machine that was used to build the compound machine.

   _____

   _____

Name _____ Date _____

Energy

Write the word or words that best complete each sentence in the spaces below. Words may be used only once.

amplitude	parallel circuit	static electricity
current electricity	prism	transparent
echo	radiation	
insulator	refracts	

1. The amount of energy in a sound wave is called its
   _____amplitude_____ .

2. When sound is reflected off a surface it is called a(n)
   _____echo_____ .

3. Energy travels from the Sun to Earth by ___radiation___ .

4. Water and clear plastic wrap are ___transparent___ .

5. In a(n) ___parallel circuit___ electricity can flow
   through more than one path.

6. A(n) ___prism___ can separate white light into
   bands of colored light.

7. A balloon can stick to a wall because of ___static electricity___ .

8. Light ___refracts___ , or bends, when it travels
   through materials like glass.

9. A(n) ___insulator___ does not transfer heat.

10. The flow of electrical charges through a circuit is
    called ___current electricity___ .

Circle the letter of the best answer for each question.

11. Sound travels <u>most</u> quickly through

   A gases, such as air.

   B liquids, such as water.

   C solids, such as wood.

   D a vacuum, which is empty.

12. Conduction occurs when energy

   A is transferred between two objects that are touching.

   B is transferred through liquids or gases.

   C is transferred over a great distance.

   D flows from a cold object to a warm one.

13. A concave lens causes light to

   A bend outward from the center so light rays spread apart.

   B bend toward the center so light rays meet.

   C cause objects to appear bigger.

   D change color when it passes through its center.

14. Electric motors change

   A mechanical energy into electrical energy.

   B chemical energy into mechanical energy.

   C electrical energy into mechanical energy.

   D mechanical energy into magnetic energy.

15. A set of angled blades attached to a shaft that provides mechanical energy for a generator is called

   A a gear.

   B a turbine.

   C a motor.

   D an electromagnet.

Name _____ Date _____

Answer the following questions.

16. **Interpret Data** Study the chart below.

Step 1	sound waves make eardrum vibrate	hammer, anvil, stirrup begin to move
Step 2	moving bones pass vibrations into inner ear	fluid in a coiled tube in the inner ear begins to vibrate
Step 3	tiny hairs inside the tube vibrate	nerves are stimulated and bring sound messages to the brain

What would be one possible effect of a person not having enough fluid in the coiled tube of the inner ear?

The person might not be able to hear properly.

_____

_____

17. **Classify** Describe one way in which sound and light are the same and one way in which they are different.

Sound and light both travel in waves. Sound travels more quickly

through solids, while light travels more slowly through denser

materials. Therefore, light travels faster through the air than sound.

_____

18. **Experiment** Several magnets are sitting on a desk. One magnet has its north and south poles marked. The rest do not. Design an experiment in which the poles of the unmarked magnets can be identified.

Answers will vary but may include: The magnet with the marked

north pole can be brought near the poles of the unmarked

magnets. If the poles attract, the unmarked magnet's pole is south;

if the poles repel, it is north. The process can be repeated to

identify the other magnets' poles.

_____

Answer the following questions.

19. **Critical Thinking** Why do you think most scientists use the Celsius scale to measure temperature? Explain.

   Answers will vary but may include: Scientists probably use the

   Celsius scale because it is easier to work with. The freezing point

   for water is 0°C and the boiling point is 100°C.

20. **Thinking Like a Scientist** How is a series circuit different from a parallel circuit? Give an example to illustrate the difference.

   Answers will vary but may include: In a series circuit, electrical

   current flows in one path. If one part of the circuit is removed

   or disconnected, the circuit is open and the flow of electricity

   is interrupted. In a parallel circuit, the circuit remains closed

   even when one part is removed or turned off because there is

   an alternate path for the current to flow. An example of a series

   circuit could be a string of holiday lights. An example of a parallel

   circuit could be the electrical outlets in homes and schools.

Name _____ Date _____

Circle the letter of the best answer for each question.

1. Heat is the flow of

   A chemical energy.

   B nuclear energy.

   C potential energy.

   D thermal energy.

2. Which of these has the fastest moving particles?

   A a lightbulb

   B a campfire

   C a piece of toast

   D a cup of hot tea

3. In order for conduction to occur, two objects must be

   A moving.

   B far apart.

   C touching.

   D made of liquid or gas.

4. Study the diagram below.

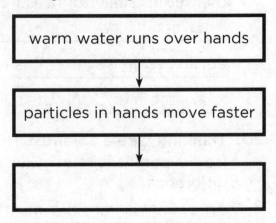

   warm water runs over hands

   ↓

   particles in hands move faster

   ↓

   Which of these belongs in the blank box?

   A hand particles gain energy

   B water particles move faster

   C hand particles move closer together

   D hand particles lose energy

Critical Thinking A glass baking dish will crack if moved from a very hot oven directly into the freezer. Why do you think this is?

Possible answer: Heat causes matter to expand, and cold causes

matter to contract. When a very hot object like glass is cooled off too

quickly, it contracts fast and cracks.

Circle the letter of the best answer for each question.

1. Humans produce sound when their vocal chords

   A echo.

   B conduct heat.

   C vibrate.

   D expand.

2. The distance from the top of one sound wave to the top of the next sound wave is called a(n)

   A pitch.

   B frequency.

   C wavelength.

   D amplitude.

3. The loudness or softness of sound is called

   A amplitude.

   B volume.

   C pitch.

   D sonar.

4. Sonar measures underwater distances by using

   A light.

   B echoes.

   C pictures.

   D heat waves.

Critical Thinking Do you think that sound would travel faster through wood or water? Explain your answer.

Possible answer: Sound would travel faster through wood. Sound is

created when particles bump into each other. Since particles in a solid

are closer together than water particles, sound travels faster through

wood.

Circle the letter of the best answer for each question.

1. Which of these statements about white light is not true?

   A It is made up of all colors.

   B It travels in waves.

   C It contains only yellow and blue.

   D It can be separated with a prism.

2. Sunburn is caused by

   A white light.

   B x-rays.

   C ultraviolet light.

   D infrared light.

3. A convex lens

   A reflects light.

   B bends light rays toward the center.

   C spreads light rays apart.

   D makes objects look smaller.

4. Study the diagram below.

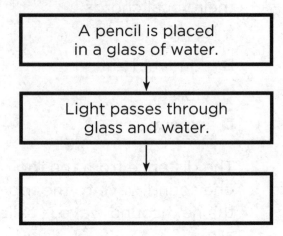

Which of these belongs in the blank box?

   A The pencil appears to be many different colors.

   B The pencil appears to be straight.

   C The pencil appears to be upside down.

   D The pencil appears to be broken.

Critical Thinking How are convex mirrors different from convex lenses?

A convex mirror reflects light and makes things appear smaller. A

convex lens does not reflect light and makes objects appear larger.

Circle the letter of the best answer for each question.

1. What causes clothes from a dryer to stick together?

   A the attraction of positive and negative charges

   B the attraction of two positive charges

   C the attraction of two negative charges

   D positive and negative charges that repel one another

2. Lightning is a type of

   A negative charges.

   B static electricity.

   C current electricity.

   D short circuit.

3. A switch in a circuit

   A acts as an insulator.

   B absorbs electricity.

   C allows or stops the flow of electricity.

   D keeps the flow of electricity at a safe level.

4. Study the diagram below.

   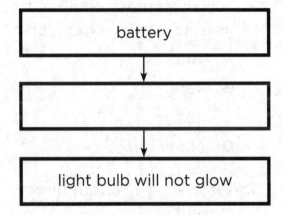

   Which of these belongs in the blank box?

   A light bulb glows

   B lightning is created

   C a spark is created

   D broken path

Critical Thinking Why do you think electrical wires are covered in rubber or plastic?

Possible answer: Rubber or plastic does not conduct electricity, so

when it is wrapped around wires it keeps people from getting a shock.

Circle the letter of the best answer for each question.

1. A motor is something that

   (A) changes electrical energy to mechanical energy.

   B changes mechanical energy to electrical energy.

   C controls the strength of an electromagnet.

   D attracts metal objects.

2. Which of these could <u>not</u> be used to power a generator?

   A wind

   B water

   C steam

   (D) aluminum

3. Which is true about direct current?

   A It runs in two directions.

   (B) It runs in one direction.

   C It is used to power home electrical outlets.

   D It is no longer used.

4. Study the diagram below.

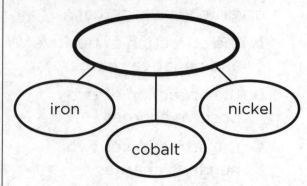

Which of these belongs in the blank oval?

   (A) Materials that Magnets Attract

   B Types of Magnets

   C Parts of an Electromagnet

   D Materials Not Attracted to Magnets

Critical Thinking Why do you think a clothes dryer requires a 220-volt outlet, while a toaster only requires a 110-volt outlet?

Possible answer: A clothes dryer uses more energy, so it requires a

stronger power source. A toaster does not use as much energy and can

be powered with a 110-volt outlet.

Energy

Write the word or words that best complete each sentence in the spaces below. Words may be used only once.

circuit	heat	poles	vibrates
discharge	insulators	sound wave	
electromagnet	magnetic field	translucent	

1. When a string moves back and forth quickly it ____vibrates____ .

2. Materials such as wool or fleece are good ____insulators____ .

3. A(n)____sound wave____ moves away from a vibrating object in all directions.

4. The flow of thermal energy from one object to another is called ____heat____ .

5. The fast movement of a charge that can cause a shock is called a(n) ____discharge____ .

6. A wire wrapped around a metal core forms a(n) ____electromagnet____ .

7. Materials that are ____translucent____ allow light to pass through them.

8. For electricity to flow, it must follow a complete path called a(n) ____circuit____ .

9. Every magnet has its own invisible ____magnetic field____ .

10. A magnet is strongest at its ____poles____ .

Circle the letter of the best answer for each question.

11. When the flow of an electric current only travels in one direction it is called

   A. voltage.

   B electromagnetic current.

   C alternating current.

   (D) direct current.

12. Which of the following are used for reading glasses to make objects nearby appear bigger?

   A concave lenses

   (B) convex lenses

   C glass prisms

   D mirrored lenses

13. How are a fuse box and a circuit breaker similar?

   A They both control static discharges.

   B They both control if a circuit is series or parallel.

   C They both determine how much current should flow.

   (D) They both open a circuit if too much current is flowing.

14. Which of the following sounds has the highest pitch?

   A a dog barking

   B a person walking on the beach

   (C) a referee blowing a whistle

   D an audience clapping

15. Which process is necessary to make a cup of hot tea?

   A refraction

   B radiation

   C contraction

   (D) convection

Answer the following questions.

16. **Interpret Data** A student rubs a balloon with wool and
sticks the balloon to a wall. Look at the chart below.

	Balloon's Charge	**Wall's Charge**
1. balloon is blown up	neutral	neutral
2. balloon is rubbed with wool	negative	neutral
3. balloon is stuck on wall	negative	positive charges move to surface of wall
4. balloon falls off the wall	neutral	neutral

Why does the balloon fall off the wall? Explain and
complete the chart.

The charges move between the balloon and the wall. Both the

balloon and the wall eventually return to a neutral state and the

balloon falls off the wall.

17. **Classify** How are reflection and refraction different?
Give an example of each.

Answers will vary but may include: Reflection is when light

bounces off a surface; refraction is when light bends as it travels

through a substance. Mirrors reflect light; water refracts light.

18. **Experiment** Three identical glasses of hot water have
three objects placed in them—a silver spoon, a plastic
spoon, and a wooden spoon.

These are allowed to sit for five minutes. Which of the
spoons will be the warmest after five minutes? Why?

The silver spoon will be the warmest because metal is a good

conductor of heat. Plastic and wood are not good conductors.

Answer the following questions.

19. Critical Thinking Explain why an American scientist might need a thermometer that measures temperatures in both Celsius and Fahrenheit.

Answers will vary but may include: Scientists usually measure temperature in Celsius, which is easier to work with. Most of the world measures temperature in Celsius, but people in the United States use Fahrenheit. Depending on the scientist's purposes, she might want to report her findings in Fahrenheit for an American audience.

20. Thinking Like a Scientist A scientist needs to set up an experiment that uses electricity. The scientist needs materials that will conduct electricity, materials that are insulators, and a device that helps prevent short circuits. What materials should the scientist use for a successful and safe experiment?

Answers will vary, but may include: Metals such as copper can be used to conduct electricity; glass, rubber, and plastic are good insulators; surge protectors, fuses, and circuit breakers prevent short circuits.

Energy

Magnet Properties

Objective: Students will first experiment with permanent magnets throughout the classroom and record their observations. They will then build an electromagnet with the nail, wire, and battery and experiment with reversing the direction of the electric current, and record their observations.

Scoring Rubric

4 points Student completes all of the steps in the experiment and accurately records all observations. Student tests permanent magnet throughout the classroom, and then successfully builds an electromagnet. Answers to questions in Analyze the Results are correct.

3 points Student completes most of the steps in the experiment and accurately records all observations. Student tests permanent magnet throughout the classroom and then builds an electromagnet. Answers to questions in Analyze the Results are mostly correct.

2 points Student completes some of the steps in the experiment and records some observations. Student may test the permanent magnet throughout the classroom or build an electromagnet, but not both. Answers to questions in Analyze the Results are mostly incorrect.

1 point Student attempts the assignment but does not complete either part of the experiment and the answers to the questions are incomplete.

Materials

- permanent magnets (2)
- insulated copper wire (1 meter long)
- wire strippers
- iron or steel nail
- Size D battery
- electrical tape

Magnet Properties

Experiment

Use the materials provided by your teacher to experiment with magnets. Walk around the classroom and test to see which objects the magnets attract. Record your observations.

Now build your own electromagnet. Wrap the copper wire tightly around the nail, leaving several inches of loose wire on either end of the nail. Then strip about an inch of the coating from each end of the wire using the wire strippers. Ask your teacher for help, if necessary. Next, use electrical tape to attach one end of the wire to one end of the battery, and the other end of the wire to the other end of the battery. Point one end of the nail to different metal objects and see if your magnet works. Find the poles of the electromagnet when the wires are connected in one direction. Then reverse the wires and test it again. Record your observations.

Analyze the Results

1. What are some similarities and differences between electromagnets and permanent magnets? How does your experiment demonstrate this?

   Both kinds of magnets attract and hold different kinds of metals.

   Permanent magnets exhibit a magnetic force continuously, while

   electromagnets can be turned on and off. When the electromagnet

   is turned off, it no longer attracts metal.

2. What did you observe when you reversed the direction of the electric current in the electromagnet? Why did this happen?

   Answers may vary but may include: The poles of the

   electromagnet were reversed. The magnetic force is reversed

   when the direction of the electric current is reversed.